Ben Stacy Jerrik (Ed.)

Wäschberg

Ben Stacy Jerrik (Ed.)

Wäschberg

Regierungsbezirk, Gemeinde, Municipalities of Germany, Miesbach, Bad Wiessee

Part Press

Imprint

Publisher:
Part Press is a trademark of
International Book Market Service Ltd., 17 Rue Meldrum, Beau Bassin, 1713-01 Mauritius
Email: info@bookmarketservice.com
Website: www.bookmarketservice.com

Published in 2013

Printed in: U.S.A., U.K., Germany. This book was not produced in Mauritius.

ISBN: 978-613-8-67221-0

Contents

Articles

References

Waltersdorf, Thuringia

Waltersdorf	
 Waltersdorf	
Coordinates	50°49′53″N 11°44′44″E
Administration	
Country	Germany
State	Thuringia
District	Saale-Holzland-Kreis
Municipal assoc.	Hügelland/Täler
Mayor	Helmut Putze
Basic statistics	
Area	4.61 km^2 (**unknown operator: u'strong'** sq mi)
Elevation	215 m (705 ft)
Population	177 *(31 December 2006)*
- Density	38 /km^2 (99 /sq mi)
Other information	
Time zone	CET/CEST (UTC+1/+2)
Licence plate	SHK
Postal code	07646
Area code	036428
Website	www.huegelland-taeler.de [1]

Waltersdorf is a municipality in the district Saale-Holzland, in Thuringia, Germany.

References

[1] http://www.huegelland-taeler.de

Gemeinde

Gemeinde (plural: ***Gemeinden***) is a German word for borough, commune, community, township, municipality, or in religious contexts, a parish or congregation (sometimes, *Kirchengemeinde*). [1]

Gemeinde also means, approximately, the parish or the community. This term in its theological usage, means more than just a community or parish. It has no English equivalent except among Mormons, who call it a ward. It encompasses the meaning of parish assembly (as a group entity) but also the congregation as a *koinonia* (Greek) or fellowship of believers.

See also

- Gemeente
- Gmina
- Municipalities of Germany
- Municipalities of Austria
- Communes of Switzerland
- Municipalities of South Tyrol

Notes

[1] "Travlang's German-English On-line Dictionary" (result for "Gemeinde"), Travlang (travlang.com), 2000, webpage: travlang-Gemeinde (http://dictionaries.travlang.com/GermanEnglish/dict.cgi?query=Gemeinde&max=50).

Municipalities of Germany

Municipalities (*Gemeinde*) are the official lowest level of territorial division **in Germany**. This may be the fourth level of territorial division in Germany, apart from those states which include *Regierungsbezirke* (singular: *Regierungsbezirk*), where municipalities then become the fifth level. The highest degree of autonomy for a municipality in Germany is when it is separate from the next higher division. It is then *kreisfrei* and generally referred to as having *city status*. This can be the case regardless of its size. Such a municipality then is the third or forth level of territorial division. Yet, many smaller municipalities in Germany have lost this city status in various administrative reforms in the last 40 years when they were amalgamated with the next higher division (Kreis). In some states they did retain a higher measure of autonomy that is than that of the other municipalities of the Kreis (e. g. *Große Kreisstadt*). That a municipality may refer to itself as a *city* or officially call itself *City of ...* has no legal significance.

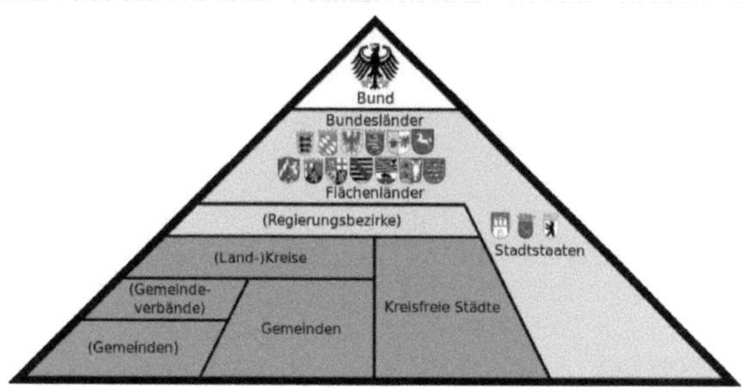

The "pyramid" of German administrative subdivisions

Overview

With more than 3,400,000 inhabitants, the most populated municipality of Germany is the city of Berlin; and the least populated is Wiedenborstel (8 inhabitants in 2010), in Schleswig-Holstein. The cities of Aachen and Saarbrücken have a special status, which is why the numbers in the respective states North Rhine-Westphalia and Saarland appear in brackets.

Municipalities per federal state

List updated at August 1, 2009.

Federal state	Municipalities	M. with city status	Average nr. of inhabitants[1]	Average surface (km²)[2]	List (Cities, Towns, Municipalities)
Baden-Württemberg	1,101	9	9,764	32.41	C, T, M
Bavaria	2,056	25	6,090	33.03	C, T, M
Berlin	1	1	3,416,255	891.02	Berlin
Brandenburg	419	4	6,052	70.36	C, T, M
Bremen	2	2	331,541	202.14	Bremen, Bremerhaven
Hamburg	1	1	1,770,629	755.16	Hamburg
Hesse	426	5	14,255	48.80	C, T, M
Lower Saxony	1,022	9	7,800	45.25	C, T, M
Mecklenburg-Vorpommern	818	2	2,053	28.34	C, T, M
North Rhine-Westphalia	236	22 (23)	45,446	86.08	C, T, M
Rhineland-Palatinate	2,306	12	1,754	8.61	C, T, M
Saarland	52	(1)	19,935	49.40	C, T, M
Saxony	491	3	8,595	37.51	C, T, M
Saxony-Anhalt	851	3	2,835	24.03	C, T, M
Schleswig-Holstein	1,116	4	2,542	14.07	C, T, M
Thuringia	955	6	2,397	16.93	C, T, M
Germany	**12,013**	**108 (110)**	**6,844**	**29.35**	**C, T, M**

References

[1] Source: Statistisches Bundesamt (as of December 31, 2007)
[2] Source: Statistisches Bundesamt (as of December 31, 2006), for the computation less the municipality-free areas

See also

- List of municipalities in Germany

Weilheim an der Teck

Weilheim an der Teck	

 Weilheim an der Teck	
Coordinates	48°36′54″N 9°32′19″E
Administration	
Country	Germany
State	Baden-Württemberg
Admin. region	Stuttgart
District	Esslingen
Town subdivisions	3
Mayor	Johannes Züfle
Basic statistics	
Area	26.51 km^2 (**unknown operator: u'strong'** sq mi)
Elevation	385 m (1263 ft)

Population	9640 *(31 December 2006)*
- Density	364 /km^2 (942 /sq mi)
Other information	
Time zone	CET/CEST (UTC+1/+2)
Licence plate	ES
Postal codes	73231–73235
Area code	07023
Website	www.weilheim-teck.de [1]

Weilheim an der Teck is a town in the district of Esslingen in Baden-Württemberg in southern Germany. It is situated 7 km southeast of Kirchheim unter Teck, and 13 km southwest of Göppingen.

References

[1] http://www.weilheim-teck.de/

This article incorporates information from the German Wikipedia.

Regierungsbezirk

The **Regierungsbezirk** pronounced [ʁeˈɡiːʁʊŋsbəˌtsɪʁk] (English: Government District) is a subdivision in 4 of the 16 federal states in Germany. It is governed by a *Bezirksregierung* (English: District Government) and led by a *Regierungspräsident* (English: Government President). The *Regierungsbezirke* do not pass any legislation. They are mostly concerned with administrative decisions on a regional level for the *Districts* covered.[1]

Regierungsbezirke as from 1 August 2008. In the map there are also shown the former RBs of Lower Saxony, Rhineland-Palatinate and Saxony-Anhalt and Saxony

History

The first *Regierungsbezirke* were created by the Kingdom of Prussia, which divided its provinces into 25 *Regierungsbezirke* in 1808/1816. The *Regierungsbezirke* of North Rhine-Westphalia are in direct continuation of those created in 1815. Other states of the German Empire created similar entities, named *Kreishauptmannschaft* (in Saxony) or *Kreis* (in Bavaria and Württemberg) (not to be confused with the *Kreis* or *Landkreis* today). During the Third Reich the naming was unified to *Regierungsbezirk*.

Regierungsbezirke by state

Today only 4 of the 16 states are divided into *Regierungsbezirke*, all others are directly divided into districts. Currently, those four states are divided into a total of 19 *Regierungsbezirke*, ranging in population from 5,255,000 (Düsseldorf) to 1,065,000 (Gießen):

- Baden-Württemberg: Freiburg, Karlsruhe, Stuttgart, Tübingen
- Bavaria: Upper Bavaria, Lower Bavaria, Upper Palatinate, Upper Franconia, Middle Franconia, Lower Franconia, Swabia
- Hesse: Darmstadt, Gießen, Kassel
- North Rhine-Westphalia: Arnsberg, Cologne, Detmold, Düsseldorf, Münster

Former states with Regierungsbezirke

On January 1, 2000 Rhineland-Palatinate disbanded its three *Regierungsbezirke* Koblenz, Rheinhessen-Pfalz and Trier - the employees and assets of the three *Bezirksregierungen* were converted into three public authorities responsible for the whole state, each covering a part of the former responsibilities of the *Bezirksregierung*.

On January 1, 2004, Saxony-Anhalt disbanded its three *Regierungsbezirke*: Dessau, Halle and Magdeburg. The responsibilities are now covered by a *Landesverwaltungsamt* with three offices at the former seats of the *Bezirksregierungen*.

At the foundation of Lower Saxony in 1946 by the merger of the three former Free States of Brunswick, Oldenburg, Schaumburg-Lippe and the former Prussian province of Hanover the former two states became *Verwaltungsbezirke* (roughly administrative regions of extended competence) within Lower Saxony besides the less autonomous Prussian-style *Regierungsbezirke* comprising the former Province of Hanover and the tiny Schaumburg-Lippe. These differences were levelled on 1 January 1978, when four territorially redeployed *Regierungsbezirke* replaced the two *Verwaltungsbezirke* and the old six *Regierungsbezirke*: Brunswick and Oldenburg as well as Aurich, Hanover (remaining mostly the same), Hildesheim, Lüneburg (old), Osnabrück and Stade. On 1 January 2005, Lower Saxony disbanded its remaining four *Regierungsbezirke*: Brunswick, Hanover, Lüneburg, and Weser-Ems.

On August 1, 2008, Saxony restructured its districts (*Landkreise*) and changed the name of its *Regierungsbezirke* to *Direktionsbezirke*. This was necessary because one of the new districts did not fit with the borders of the old *Regierungsbezirke* and some responsibilities are now covered by the districts. The *Direktionsbezirke* are still named *Chemnitz*, *Dresden* and *Leipzig*. As of March 1, 2012, the *Direktionsbezirke* were merged into one *Landesdirektion*.

References

[1] Regional Governments in France, Germany, Poland and The Netherlands (HTML version of PowerPoint presentation) (http://google.com/search?q=cache:I1oTRcSOTJ8J:www.hanse-passage.net/hansepassage/hpcms/uploads/_presentation%20analysis.ppt+"regierungsbezirk+is"&hl=en&ct=clnk&cd=18&lr=lang_en) - Cachet, A (coordinator), Erasmus University, Rotterdam

Baden-Württemberg

Baden-Württemberg	
— State of Germany —	
 Flag	
Coordinates: 48°32′16″N 9°2′28″E	
Country	Germany
Capital	Stuttgart
Government	
• Minister-President	Winfried Kretschmann (Greens)
• Governing parties	Greens / SPD
• Votes in Bundesrat	6 (of 69)
Area	
• Total	35751.65 km^2 (**unknown operator: u'strong'** sq mi)
Population (2011-11-30)[1]	
• Total	10789000
• Density	**unknown operator: u'strong'**/km^2 (**unknown operator: u'strong'**/sq mi)
Time zone	CET (UTC+1)
• Summer (DST)	CEST (UTC+2)
ISO 3166 code	DE-BW
GDP/ Nominal	€376.28 billion (2011) [2]
NUTS Region	DE1
Website	www.baden-wuerttemberg.de [3]

Baden-Württemberg (German pronunciation: [ˈbaːdən ˈvʏʁ̞təmˌbɛʁ̞k]) is one of the 16 states of Germany. Baden-Württemberg is in the southwestern part of the country to the east of the Upper Rhine, and is the third largest of Germany's sixteen states in terms of both area and population, with an area of 35742 square kilometres (**unknown operator: u'strong'** sq mi) and 10.7 million inhabitants.[4] The state capital is Stuttgart.

The sobriquet *Ländle* ("small land" in the local dialect) is sometimes used as a synonym for the Swabian part of Baden-Württemberg.

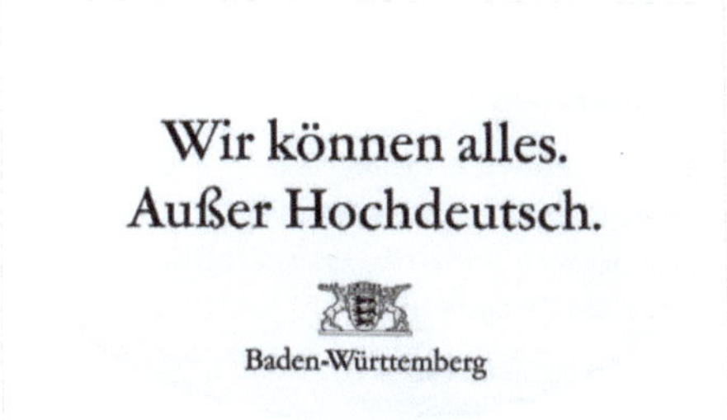

A campaign sticker, translated, "We can [do] anything. Except [speak] Standard German." This is an allusion to Baden-Württemberg being one of the principal centres for innovation in Germany and having many inhabitants with distinctive dialects.

History

The area used to be covered by the historical states of Baden, including the former Prussian Hohenzollern, and Württemberg, part of the region of Swabia.[5]

In the first century AD, Württemberg was occupied by the Romans, who defended their position there by constructing a limes (fortified boundary zone). Early on in the third century, the Alemanni drove the Romans beyond the Rhine and the Danube, but in their turn they succumbed to the Franks under Clovis I, the decisive battle taking place in 496. It later became part of the Holy Roman Empire.

After World War II, Allied forces established three federal states: Württemberg-Hohenzollern, Baden (both occupied by France), and Württemberg-Baden (US-occupied). In 1949, these three states became founding members of the Federal Republic of Germany. Article 118 of the new German constitution, however, had already prepared a procedure for those states to merge. After a referendum held on 16 December 1951, Württemberg-Baden, Württemberg-Hohenzollern and Baden voted in favor of a merger.[4] Baden-Württemberg officially became a state on 25 April 1952.[4]

Geography

Baden-Württemberg shares borders with France (departments of Haut-Rhin and Bas-Rhin), Switzerland (cantons of Basel-Landschaft, Basel-Stadt, Aargau, Zurich, Schaffhausen and Thurgau) and the states of Rhineland Palatinate, Hesse and Bavaria.[4]

Most of the major cities of Baden-Württemberg straddle the banks of the Neckar River, which runs downstream (from southwest to the centre, then northwest) through the state first past Tübingen, then Stuttgart, Heilbronn, Heidelberg, and Mannheim).

The Rhine (German: *Rhein*) forms the western border as well as large portions of the southern border. The Black Forest (Schwarzwald), the main mountain range of the state, rises east of the Rhine valley. The high plateau of the Swabian Alb, between the Neckar, the Black Forest and the Danube, is an important European watershed. Baden-Württemberg shares Lake Constance (Bodensee; the border with Switzerland is the middle of the lake) with Switzerland and it shares the foothills of the Alps (known as the Allgäu) with Bavaria and the Austrian Vorarlberg (but Baden-Württemberg does not border Austria over land).

The Danube (Donau) river has its source in Baden-Württemberg near the town of Donaueschingen, in a place called Furtwangen in the Black Forest.

Government

Administration

Baden-Württemberg is divided into 35 districts (Landkreise) and 9 independent cities (Stadtkreise), both grouped into the four Administrative Districts (*Regierungsbezirke*) of Freiburg, Karlsruhe, Stuttgart, and Tübingen.

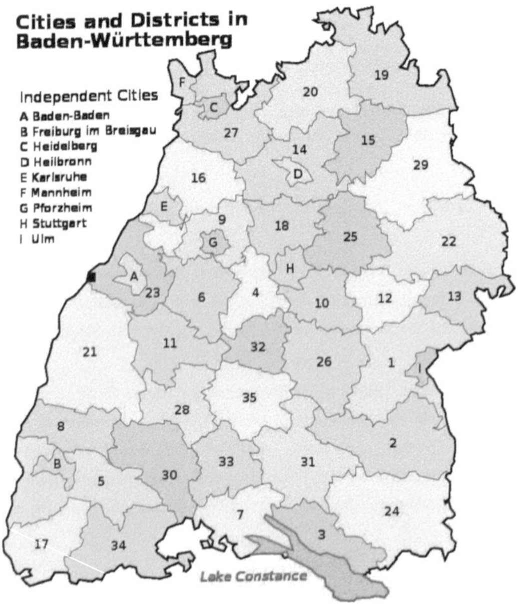

Map

Stuttgart

Karlsruhe

Heidelberg

Freiburg

Mannheim

Ulm

Baden-Baden

Pforzheim

- Alb-Donau
- Biberach
- Bodensee
- Böblingen
- Breisgau-Hochschwarzwald
- Calw
- Konstanz *(Constance)*
- Emmendingen
- Enzkreis
- Esslingen
- Freudenstadt
- Göppingen
- Heidenheim
- Heilbronn
- Hohenlohe
- Karlsruhe
- Lörrach
- Ludwigsburg
- Main-Tauber
- Neckar-Odenwald-Kreis
- Ortenaukreis
- Ostalbkreis
- Rastatt
- Ravensburg
- Rems-Murr-Kreis
- Reutlingen
- Rhein-Neckar-Kreis
- Rottweil
- Schwäbisch Hall
- Schwarzwald-Baar-Kreis
- Sigmaringen
- Tübingen
- Tuttlingen
- Waldshut
- Zollernalbkreis

Furthermore there are nine independent cities, which do not belong to any district:

Code	City (*stadtkreise*)	Area (km^2)	Population 1997	Population 2007	Region (*regierungs-bezirk*)
A	Baden-Baden	140.18	52,672	54,853	Karlsruhe
B	Freiburg im Breisgau	153.06	200,519	219,430	Freiburg
C	Heidelberg	108.83	139,941	145,311	Karlsruhe
D	Heilbronn	99.88	120,987	121,627	Stuttgart
E	Karlsruhe	173.46	276,571	288,917	Karlsruhe
F	Mannheim	144.96	310,475	309,795	Karlsruhe
G	Pforzheim	98.02	118,079	119,423	Karlsruhe
H	Stuttgart	207.35	585,274	597,176	Stuttgart
I	Ulm	118.69	115,628	121,434	Tübingen

Politics

The state parliament of Baden-Württemberg is the **Landtag** (Eng. state assembly). The politics of Baden-Württemberg have traditionally been dominated by the conservative Christian Democratic Union of Germany (CDU), who until 2011 had led all but one government since the establishment of the state in 1952. In the *Landtag* elections held on 27 March 2011 voters replaced the Christian Democrats and centre-right Free Democrats coalition by a Greens-led alliance with the Social Democrats which secured a four seat majority in the state parliament.
From 1992 to 2001, the anti-immigration Republicans party had seats in the *Landtag*.[6]

Further supreme institutions of the state

The Baden-Württemberg General Auditing Office acts as an independent body to monitor the correct use of public funds by public offices.[7]

Economy

Although Baden-Württemberg has relatively few natural resources compared to other regions of Germany,[4] the state is among the most prosperous[5] and wealthiest regions in Europe with a generally low unemployment rate historically. A number of well-known enterprises are headquartered in the state, for example Daimler AG, Porsche, Robert Bosch GmbH (automobile industry), Carl Zeiss AG (optics), and SAP AG (largest software enterprise in Europe). In spite of this, Baden-Württemberg's economy is dominated by small and medium-sized enterprises. Although poor in workable natural resources (formerly lead, zinc, iron, silver, copper and salts) and still rural in many areas, the region is heavily industrialized. In 2003, there were almost 8,800 manufacturing enterprises with more than 20 employees, but only 384 with more than 500. The latter category accounts for 43% of the 1.2 million persons employed in industry. The *Mittelstand* or mid-sized company is the backbone of the Baden-Württemberg economy.[8] Medium-sized businesses and a tradition of branching out into different industrial sectors have ensured specialization over a wide range. A fifth of the "old" Federal Republic's industrial gross value added is generated by Baden-Württemberg. Turnover for manufacturing in 2003 exceeded 240,000 million, 43% of which came from exports. The region depends to some extent on global economic developments, though the great adaptability of the region's economy has generally helped it through crises. Half of the employees in the manufacturing industry are in mechanical and electrical engineering and automobile construction. This is also where the largest enterprises are to be found. The importance of the precision mechanics industry also extends beyond the region's borders, as does that of the optical, clock making, toy, metallurgy and electronics industries. The textile industry, which formerly dominated much of the region, has now all but disappeared from Baden-Württemberg. Research and development (R&D) is funded jointly by the State and industry. In 2001, more than a fifth of the 100,000 or so persons working in R&D in Germany were located in Baden-Württemberg, most of them in the Stuttgart area.[9] Baden-Württemberg is also one of the Four Motors for Europe.

A study performed in 2007 by the pr campaign "Initiative for New Social Market Economy" (German: *Initiative Neue Soziale Marktwirtschaft* (INSM)) and the trade newspaper "Wirtschaftswoche" awarded Baden-Württemberg for being the "economically most successful and most dynamic state" among the 16 states.

Tourism

Baden-Württemberg is a popular holiday destination. Main sights include the capital and biggest city, Stuttgart, modern and historic at the same time, with its urban architecture and atmosphere (and famously, its inner city parks and historic Wilhelma zoo), its castles (such as Castle Solitude), its (car and art) museums as well as a rich cultural programme (theatre, opera) and mineral spring baths in Stuttgart-Bad Cannstatt (also the site of a Roman Castra); it is the only major city in Germany with vineyards in an urban territory.

The residential (court) towns of Ludwigsburg and Karlsruhe, the spas and casino of luxurious Baden-Baden, the medieval architecture of Ulm (Ulm Münster is the tallest church in the world), the vibrant, young, but traditional university towns of Heidelberg and Tübingen with their old castles looking out above the river Neckar, are popular smaller towns. Sites of former monasteries such as the ones on Reichenau Island and at Maulbronn (both World Heritage Sites) as well as Bebenhausen Abbey are to be found. Baden-Württemberg also boasts rich old Free Imperial Cities such as Biberach, Esslingen am Neckar, Heilbronn, Ravensburg, Reutlingen and Schwäbisch Hall, as well as the southernmost and sunniest city of Germany, Freiburg, close to Alsace and Switzerland, being an ideal base for exploring the heights of the nearby Black Forest (e.g. for skiing in winter or for hiking in summer) with its traditional villages and the surrounding wine country of the Rhine Valley of South Baden.[5]

The countryside of the lush Upper Neckar valley (where Rottweil is famous for its carnival (Fastnacht)) and the pristine Danube valley Swabian Alb (with Hohenzollern Castle), as well as the largely pristine Swabian Forest, the Rhine Valley and Lake Constance (German: *Bodensee*), where all kinds of water sports are popular, with the former Imperial, today border town of Konstanz (where the Council of Constance took place), the Neolithic and Bronze Age village at Unteruhldingen, the flower island of Mainau, and the hometown of the Zeppelin, Friedrichshafen a.o., are especially popular for outdoor activities in the summer months.[5]

In spring and autumn (April/May and September/October), beer festivals (fun fairs) are taking place at the Cannstatter Wasen in Stuttgart; the one in the autumn, the Cannstatter Volksfest, is the second biggest such festival in the world after the Munich Oktoberfest. In late November/early December, Christmas markets are a tourist magnet in all major towns, with the biggest one in Stuttgart, lasting for the three weeks prior to Christmas.

The Bertha Benz Memorial Route is a 194 km signposted scenic route from Mannheim via Heidelberg and Wiesloch to Pforzheim and back, which follows the route of the world's first long-distance journey by automobile which Bertha Benz undertook in August 1888.

Companies owned by Baden-Württemberg

Company	Industry	Percentage owned	Source
EnBW	Energy industry	45%	[10]
Badische Staatsbrauerei Rothaus	Beverage industry	100%	

Education

Baden-Württemberg is home to some of the oldest, most renowned and prestigious universities in Germany, such as the universities of Heidelberg, Freiburg and Tübingen. It also contains three of the eleven German 'excellence universities' (Heidelberg, Tübingen and Konstanz and formerly Freiburg and Karlsruhe).

University of Karlsruhe, since 2009: Karlsruhe Institute of Technology

Other university towns are Mannheim and Ulm. Furthermore, two universities are located in the state capital Stuttgart, the University of Hohenheim and the University of Stuttgart. Ludwigsburg is home to the renowned national film school Filmakademie Baden-Württemberg (Film Academy Baden-Wuerttemberg). The private International University in Germany is situated in Bruchsal. There is another private university, located in Friedrichshafen, Zeppelin University.

Furthermore, there are more than a dozen Fachhochschulen, i.e. universities of applied sciences, as well as *Pädagogische Hochschulen*, i.e. teacher training colleges, and other institutions of tertiary education in Baden-Württemberg (a.o. in Aalen, Biberach an der Riss, Esslingen, Karlsruhe, Ludwigsburg, Nürtingen, Pforzheim, Ravensburg-Weingarten, Reutlingen, several in Stuttgart, Schwäbisch Hall).

The state has the highest density of academic institutions of any territorial state (i.e. excluding Berlin and Hamburg) in Germany.

Dialects

Two dialect groups of German are spoken in Baden-Württemberg in various variants: Alemannic and Franconian dialects. In central and southern Württemberg, the Alemannic dialect of Swabian is spoken (slightly differing even within the area, e.g. between Upper Swabia, the Swabian Alb and the central Neckar Valley of the Stuttgart region). In South Baden, the local dialects are Low Alemannic and High Alemannic (i.e. variants of what is also Swiss German). In the northern part of Baden, i.e. the former *Kurpfalz* (Electorate of the Palatinate) with the former capitals of Heidelberg and Mannheim, the idiom is Rhine Franconian (i.e. Palatinate German), while in the Northeast East Franconian is spoken.

The same or similar Alemannic dialects are also spoken in the neighbouring regions, especially in Bavarian Swabia, Alsace (Alsatian), German-speaking Switzerland (Swiss German) and the Austrian Vorarlberg, while the other Franconian dialects range from the Netherlands over the Rhineland, Lorraine and Hesse up to northern Bavaria Franconia.

Demographics

The population of Baden-Württemberg is 10,749,755 (2008), of which 5,466,966 are female and 5,282,789 are male, total population up 0.10 per cent over a year earlier. This was due to more births than deaths. In 2006, the birth rate was 8.61 per 1000, lower than that of 8.80 per 1000 in 2005. The death rate decreased from 8.80 per 1000 in 2005 to 8.60 per 1000 in 2006. In 2008, nearly 14.87 percent of the population under the age of 15, fell from 15.13 per cent over a previous year. The proportion of people aged 65 and over rose from 18.72 per cent to 18.99 per cent. Correspondingly, the median age (aged 15–64) of the population fell from 66.15 to 66.14 over the same period. The ratio of people aged under 15 and aged 65 and over to the population of working age (aged 15–64), the overall dependency ratio is 512 per 1000 in 2008. The sex ratio of total population is 0.966 male(s)/female.

Religion

While Northern and most of central Württemberg has been traditionally Lutheran–Protestant (later also Pietist) since the Reformation that was adopted there in 1534 (with its center at the famous Tübinger Stift) and the former Electorate of the Palatinate (Northwestern Baden) with its capital Heidelberg was shaped by Calvinism before being integrated into Baden, Upper Swabia, the Upper Neckar Valley up to the bishop seat of Rottenburg and Southern Baden (the Catholic archbishop has its seat in Freiburg) have traditionally been bastions of Catholicism.

Religion	%	
Roman Catholics	36.9%[11]	4.0M
Evangelical Church in Germany	33.3%[12]	3.6M
Muslims	5.6%	600 000
Buddhists	0.23%	25 000
Hindu	0.14%	15 000
Jews	0.08%	9 000
Non Religious	22.3%	2.4M

See also

- Baden, for the pre-19th century state
- History of Baden
- Württemberg and History of Württemberg
- Minister-President of Baden-Württemberg
- Origin of the coats of arms of German federal states
- Grand Duchy of Baden, for the state that existed from 1808–1918
- Republic of Baden, for the state that existed from 1918–1945
- Rulers of Baden, for a list of sovereigns and presidents
- historical "Wappen" (coat of arms)
- List of places in Baden-Württemberg
- Four Motors of Europe
- Staatsoper Stuttgart (part of "Staatstheater")
- The Baden State Theatre at Karlsruhe

Footnotes

[1] "State population" (http://www.statistik-portal.de/Statistik-Portal/de_zs01_bw.asp). *Portal of the Federal Statistics Office Germany.*. Retrieved 31 July 2012.

[2] "GDP of state" (http://www.statistik-bw.de/VolkswPreise/Landesdaten/LRtBWSjewPreise.asp). *Portal of the Baden-Württemberg Statistics Office.*. Retrieved 31 July 2012.

[3] http://www.baden-wuerttemberg.de/en/index.html

[4] "Our State" (http://www.baden-wuerttemberg.de/en/Our_State/86236.html). Baden-Württemberg.. Retrieved 30 March 2011.

[5] Andrea Schulte-Peevers; Anthony Haywood, Sarah Johnstone, Jeremy Gray, Daniel (2007). *Germany* (http://books.google.com/books?id=Z5t5mZE_s5YC&pg=PA392#PPA391,M1). Lonely Planet. ISBN 978-1-74059-988-7.. Retrieved 1 February 2009.

[6] The *Republikaner*(REP): 15 from 146 MPs in 1992 and 14 from 155 MPs in 1996. "Baden-Württemberg. Results of the election from 1964–2011." (http://www.statistik.baden-wuerttemberg.de/Wahlen/Landesdaten/Landtagswahlen/LRLtWsitzvert.asp). Statistisches Landesamt Baden-Württemberg..

[7] "Government and organs of state" (http://www.baden-wuerttemberg.de/en/Government_and_organs_of_state/86233.html). Baden-Württemberg.. Retrieved 30 March 2011.

[8] Cooke, p. 84

[9] "BADEN – WÜRTTEMBERG – Economy" (http://circa.europa.eu/irc/dsis/regportraits/info/data/en/de1_eco.htm). *Eurostat.* June 2004. Archived (http://web.archive.org/web/20081223024151/http://circa.europa.eu/irc/dsis/regportraits/info/data/en/de1_eco.htm) from the original on 23 December 2008.. Retrieved 16 January 2009.

[10] Haase, Nina (30 March 2011). "Business leaders wary of Greens' state election victory" (http://www.dw-world.de/dw/article/0,,14951861,00.html). *Deutsche Welle.*. Retrieved 30 March 2011.

[11] chiesa cattolica http://www.dbk.de/imperia/md/content/kirchlichestatistik/bev-kath-l__nd-2008.pdf

[12] EKD http://www.ekd.de/download/kirchenmitglieder_2007.pdf

References

- Philip Cooke, Kevin Morgan (1998). *The Associational Economy: Firms, Regions, and Innovation.* Oxford University Press. p. 84. ISBN 978-0-19-829659-1.
- *Climate change in Baden-Württemberg: facts – impacts – perspectives* / LUBW; Baden-Württemberg, Ministerium für Umwelt, Naturschutz und Verkehr. [In collab. with Karlsruhe Institute of Technology (KIT), Institute for Meteorology and Climate Research; Süddeutsches Klimabüro. Transl.: InTra eG Fachübersetzergenossenschaft], Stuttgart: Baden-Württemberg Ministry of the Environment, Nature Conservation and Transport Karlsruhe: LUBW, 2010.

External links

- Official website (http://www.baden-wuerttemberg.de/en/A_warm_welcome_to_Baden-Wuerttemberg/88737.html)
- Baden-Württemberg (http://www.dmoz.org/Regional/Europe/Germany/States/Baden-WÃ¼rttemberg//) at the Open Directory Project
- Geographic data related to Baden-Württemberg (http://www.openstreetmap.org/browse/relation/62611) at OpenStreetMap

List of twin towns and sister cities in Germany

This list is incomplete.

This is a list of places in **Germany** having standing links to local communities in other countries, or, respectively in other parts of Germany (mostly across the former inner German border). In most cases, the association, especially when formalised by local government, is known as "town twinning" (though other terms, such as "partner towns" or "sister cities" are sometimes used instead), and while most of the places included are towns, the list also comprises villages, cities, districts, counties, etc. with similar links.

A few twin places have a (old or historical) *German name*, especially in bordering countries or areas with mixed languages.

For German-German twins there are abbreviated state names: Baden-Württemberg (**BW**), Bavaria (**BA**), Berlin (**BE**), Brandenburg (**BB**), Bremen (**BN**), Hamburg (**HB**), Hesse (**HE**), Lower Saxony (**LS**), Mecklenburg-Vorpommern (**MV**), North Rhine-Westphalia (**NRW**), Rhineland-Palatinate (**RP**), Saarland (**SL**), Saxony-Anhalt (**SA**), Saxony (**SX**), Schleswig-Holstein (**SH**) and Thuringia (**TH**).

A

- Aachen, North Rhine-Westphalia:

Liège (*Lüttich*), Belgium (1955) Reims, France (1967) Halifax, United Kingdom (1979) Toledo, Spain (1985)	Ningbo, China (1986) Naumburg (**SA**), Germany (1988) Arlington, Virginia, USA (1993) Cape Town, South Africa (1999)	Kostroma, Russia (2005) Rosh HaAyin, Israel (2007) Baltimore, Ireland (2010)

twinned with Kornelimünster/Walheim:	Montebourg, France (1972)

- Aalen, Baden-Württemberg:

Saint-Lô, France (1978) Christchurch, United Kingdom (1981)	Tatabánya, Hungary (1987) Antakya, Turkey (1995)	Cervia, Italy (2011)

- Abensberg, Bavaria:

Parga, Greece (1986)	Lonigo, Italy (1999)

- Adenau, Rhineland-Palatinate:

Sillery, France (1985)	Mellieħa, Malta	Castione della Presolana, Italy (2002)

- Ahlen, North Rhine-Westphalia:

borough of Tempelhof-Schöneberg, Berlin (**BE**), Germany (1964) Differdange (*Differdingen*), Luxembourg (1982)	Teltow (**BE**), Germany (1991)	Penzberg (**BA**), Germany (2007)

- Ahrensburg, Schleswig-Holstein:

Esplugues de Llobregat, Spain (1988) Viljandi (*Fellin*), Estonia (1989)	Ludwigslust (**BA**), Germany (1990)	Feldkirchen in Kärnten, Austria (1998)

- Alsfeld, Hesse:

New Mills, United Kingdom (1962) Nakskov, Denmark (1963)	Chaville, France (1974) Amstetten, Austria (1979)	Spišská Nová Ves ((*Zipser*) *Neu*(*en*)*dorf*), Slovakia (1992)

- Alzey, Rhineland-Palatinate:

Harpenden, United Kingdom (1963) Josselin, France (1973)	Lembeye, France (1980) Rechnitz, Austria (1981)	Kościan (*Kosten*), Poland (1990) Kamenz (**SX**), Germany (1990)

- Amberg, Bavaria:

Bad Bergzabern (**RP**), Germany (1939, renewed 1952) Schnaittenbach (**BA**), Germany (1954) Périgueux, France (1965) Kranj (*Krainburg*), Slovenia (1965)	Geretsried (**BA**), Germany (1970) Bystrzyca Kłodzka (*Habelschwerdt*), Poland (1988, renewed 2001) Siilinjärvi, Finland (1999) Desenzano, Italy (2006)	Freiberg (**SX**), Germany Trikala, Greece Ústí nad Orlicí (*Wildenschwert*), Czech Republic

- Andernach, Rhineland-Palatinate:

Saint-Amand-les-Eaux, France (1959) Dimona, Israel (1975)	Ekeren, Belgium (1979) Stockerau, Austria (1983)	Zella-Mehlis (**TH**), Germany (1990) Farnham, United Kingdom (1991)

- Anklam, Mecklenburg-Vorpommern:

Heide (**SH**), Germany Ustka (*Stolpmünde*), Poland (1995)	Limbaži (*Lemsal*), Latvia	Burlöv, Sweden

- Ansbach, Bavaria:

Bay City, Michigan, USA (1960) Anglet, France (1968)	Jingjiang, China (2004)	Fermo, Italy (2006)

- Arnsberg, North Rhine-Westphalia:

Deventer, Netherlands (1956) Bexley, United Kingdom (1971)	Alba Iulia (*Weißenburg/Karlsburg*), Romania (1974) Fos-sur-Mer, France (1982)	Olesno (*Rosenberg*), Poland (1992) Caltagirone, Italy (2011)

- Aschaffenburg, Bavaria:

Perth, United Kingdom (1956) Saint-Germain-en-Laye, France (1975)	Villavicencio, Colombia (1991)	Miskolc (*Mischkolz*), Hungary (1996)

- Augsburg, Bavaria:

Inverness, Scotland (1956) Nagahama, Japan (1959) Amagasaki, Japan (1959)	Dayton, Ohio, USA (1964) Bourges, France (1967) Liberec (*Reichenberg*), Czech Republic (2001)	Jinan, China (2004)

- Aurich, Lower Saxony:

Appingedam, Netherlands (1989)

B

- Backnang, Baden-Württemberg:

Annonay, France (1966)	Bácsalmás (*Almasch*), Hungary (1988)	Chelmsford, United Kingdom (1990)

- Bad Aibling, Bavaria:

Cavaion Veronese, Italy (1986)

- Bad Doberan, Mecklenburg-Vorpommern:

Bad Schwartau (**SH**), Germany (1989)

- Bad Dürrenberg, Saxony-Anhalt:

Encs, Hungary Caudebec-lès-Elbeuf, France	Melle (**LS**), Germany (1991)	Ciechocinek, Poland (1996)

- Bad Ems, Rhineland-Palatinate:

Cosne-sur-Loire, France (1974)	Droitwich Spa, United Kingdom (1983)	Blankenfelde-Mahlow (**BB**), Germany (1992)

- Bad Füssing, Bavaria:

Abano Terme, Italy

- Bad Gandersheim, Lower Saxony:

Skegness, United Kingdom (before 1982)	Rotselaar, Belgium (1987)

- Bad Griesbach im Rottal, Bavaria:

Friesach, Austria (2006)

- Bad Homburg, Hesse:

Bejaia, Algeria (1956-1963) Chur, Switzerland (1956) Mayrhofen, Austria (1956) Terracina, Italy (1956)	Mondorf-les-Bains (*Bad Mondorf*), Luxembourg (1956) Cabourg, France (1956) Exeter, United Kingdom (1965) Mariánské Lázně (*Marienbad*), Czech Republic (1991)	Petergof (*Peterhof*), Russia (1994) Dubrovnik, Croatia (2002) Greiz (**TH**), Germany (planned) Niš, Serbia (planned)

- Bad Kissingen, Bavaria:

Massa, Italy (1960)	Vernon, France (1960)	Eisenstadt, Austria (1978)

- Bad Königshofen, Bavaria:

Arlington, Texas, USA (1952)	Römhild (**TH**), Germany

- Bad Kötzting, Bavaria. Bad Kötzting is a founding member of the Douzelage,[1] a town twinning association of 23 towns across the European Union:[1] [2]

Altea, Spain (1991) Bellagio, Italy (1991) Bundoran, Ireland (1991) Granville, France (1991) Holstebro, Denmark (1991) Houffalize, Belgium (1991) Meerssen (*Meersen*), Netherlands (1991) Niederanven, Luxembourg (1991) Preveza, Greece (1991)	Sesimbra, Portugal (1991) Sherborne, United Kingdom (1991) Karkkila, Finland (1997) Oxelösund, Sweden (1998) Judenburg, Austria (1999) Chojna (*Königsberg in der Neumark*), Poland (2004) Kőszeg (*Güns*), Hungary (2004) Sigulda (*Segewold*), Latvia (2004) Sušice (*Schüttenhofen*), Czech Republic (2004)	Türi (*Turgel*), Estonia (2004) Zvolen (*Altsohl*), Slovakia (2007) Prienai (*Prenen*), Lithuania (2008) Marsaskala, Malta (2009) Siret (*Sereth*), Romania (2010) Agros, Cyprus (planned) Škofja Loka (*Bischoflack*), Slovenia (planned) Tryavna, Bulgaria (planned)

- Bad Kreuznach, Rhineland-Palatinate:

Bourg-en-Bresse, France (1963)	Neuruppin (**BB**), Germany (1990)

- Bad Mergentheim, Baden-Württemberg:

Digne-les-Bains, France (1962) Sainte-Marie-du-Mont, Manche, France (1967)	Fuefuki, Japan (1991, renewed 2007)	Borgomanero, Italy (2005)

- Bad Münstereifel, North Rhine-Westphalia:

Ashford, United Kingdom (1964)	Fougères, France (1967)

- Bad Nauheim, Hesse:

Oostkamp, Belgium (1985) Buxton, United Kingdom (1986)	Bad Langensalza (**TH**), Germany (1990) Chaumont, France (1992)	Manitowoc, Wisconsin, USA (not likely)

- Bad Oeynhausen, North Rhine-Westphalia:

Fismes, France (1968)	Wear Valley, United Kingdom (1977)	Inowrocław (*Inowrazlaw/Hohensalza*), Poland (1989)

- Bad Pyrmont, Lower Saxony:

Anzio, Italy (1958)	Bad Freienwalde (**BB**), Germany (1990)	Heemstede, Netherlands (2000)

- Bad Rappenau, Baden-Württemberg:

  Contrexéville, France (1982)

  Llandrindod Wells, Wales (2001)

- Bad Sobernheim, Rhineland-Palatinate:

Edelény (*Edeling*), Hungary (1997)	Louvres, France (1982)

- Bad Soden am Taunus, Hesse:

Rueil-Malmaison, France (1975) Kitzbühel, Austria (1984)	Františkovy Lázně (*Franzensbad*), Czech Republic (1992)	Yoro-cho, Japan (2004)

- Bad Vilbel, Hesse:

Huizen, Netherlands (1972-2000) Eldoret, Kenya (1982)	Glossop, United Kingdom (1985) Moulins, France (1990)	Brotterode (**TH**), Germany

- Bad Wildungen, Hesse:

Saffron Walden, England (1986) Yichun, China (1988)	Saint-Jean-de-Maurienne, France (1990)	Bad Berka (**TH**), Germany

- Bad Wurzach, Baden-Württemberg:

Wallingford, United Kingdom (1988) Luxeuil-les-Bains, France (2000)	Popielów, Opole Voivodeship (*Alt Poppelau*), Poland (2000)	St Helier, Jersey, United Kingdom (2002)

- Baden-Baden, Baden-Württemberg:

Menton, France (1961) Moncalieri, Italy (1990)	Freital (**SX**), Germany (1990) Karlovy Vary (*Karlsbad*), Czech Republic (1998)	Jalta, Ukraine (2000)

- Baesweiler, North Rhine-Westphalia:

Easington, United Kingdom (1975)	Montesson, France (1990)

- Balingen, Baden-Württemberg:

  Royan, France (1980)

- Bamberg, Bavaria:

Rodez, France (1970) Villach, Austria (1973) Bedford, United Kingdom (1977)	Esztergom (*Gran*), Hungary (1992) Prague (*Prag*), Czech Republic (1992) Feldkirchen in Kärnten, Austria (1993)	Nagaoka, Japan (1995)

- Bassum, Lower Saxony:

Fresnay-sur-Sarthe, France (1972)	Telšiai (*Telsche(n)*), Lithuania (2009)	Spilsby, United Kingdom (2010)

- Battenberg, Hesse:

Loon op Zand, Netherlands (1962) Senonches, France (1979)	Romsey, United Kingdom (1987) Litvínov (*Oberleutensdorf*), Czech Republic	Horní Jiřetín (*Obergeorgenthal*), Czech Republic

- Bautzen, Saxony:

Worms (**RP**), Germany (1990) Heidelberg (**BW**), Germany (1991)	Dreux, France (1992) Jablonec nad Nisou (*Gablonz an der Neiße*), Czech Republic (1993)	Jelenia Góra (*Hirschberg*), Poland (1993)

- Bayreuth, Bavaria:

Annecy, France (1966) Rudolstadt (**TH**), Germany (1990) State Burgenland, Austria (1990)	La Spezia, Italy (1999) 6th district of Prague, Czech Republic (2008) Tekirdağ, Turkey (2010)	Shrewsbury, United Kingdom (planned)

- Belm, Lower Saxony:

Elterlein (**SX**), Germany (2004)	Kolno (*Groß Köllen*), Poland (2006)	Englewood, Colorado, USA (2007)

- Bensheim, Hesse:

Beaune, France (1960) Amersham, United Kingdom (1977)	Mohács (*Mohatsch*), Hungary (1987) Riva del Garda, Italy (1989)	Kłodzko (*Glatz*), Poland (1996) Hostinné (*Arnau*), Czech Republic (2002)

- Bergisch Gladbach, North Rhine-Westphalia:

Bourgoin-Jallieu, France (1956) Luton, England (1956) Velsen, Netherlands (1956)	Joinville-le-Pont, France (1960) Runnymede, United Kingdom (1965) Marijampolė (*Mariampol*), Lithuania (1989)	Limassol, Cyprus (1991) Pszczyna (*Pless*), Poland (1993) Beit Jala, Palestine (2011)

- Berlin, Berlin:

Los Angeles, California, USA (1967) Paris, France (1987) Madrid, Spain (1988) Istanbul, Turkey (1989) Warsaw, Poland (1991) Moscow, Russia (1991) Budapest, Hungary (1991)	Brussels, Belgium (1992) Jakarta, Indonesia (1993) Tashkent, Uzbekistan (1993) Mexico City, Mexico (1993) Beijing, China (1994) Tokyo, Japan (1994) Buenos Aires, Argentina (1994)	Brasilia, Brazil (1994, not official) Prague, Czech Republic (1995) Windhoek, Namibia (2000) London, United Kingdom (2000) Shiraz, Iran (2007, not official)

Mannheim (**BW**), Germany (1962 by C.) Marburg-Biedenkopf District (**HE**), Germany (1964 by C.) Or Yehuda, Israel (1966 by C.) Trento, Italy (1966 by C.) Apeldoorn, Netherlands (1968 by W.) Gladsaxe, Denmark (1968 by W.) Minden (**NRW**), Germany (1968 by W.)	London Borough of Lewisham, United Kingdom (1968 by C.) London Borough of Sutton, United Kingdom (1968 by W.) Split, Croatia (1970 by W.) Bad Iburg (**LS**), Germany (1980 by C.) Karmiel, Israel (1985 by W.) Waldeck-Frankenberg District (**HE**), Germany (1988 by C.) Pechersk District Kiev, Ukraine (1991 by W.)	Rheingau-Taunus District (**HE**), Germany (1991 by W.) Forchheim District (**BA**), Germany (1991 by W.) Kulmbach District (**BA**), Germany (1991 by W.) Gagny, France (1992 by W.) Międzyrzecz (*Meseritz*), Poland (1993 by W.) Linz, Österreich (2001, 1995 by C.) District 5 Budapest (Inner City), Hungary (1998 by C.)

Wiesbaden (**HE**), Germany (1964) Bergstraße District (**HE**), Germany (1969) Ingelheim am Rhein (**RP**), Germany (1971) Porta Westfalica (**NRW**), Germany (1973)	Limburg-Weilburg District (**HE**), Germany (1980) San Rafael del Sur, Nicaragua (1986) Kiryat Yam, Israel (1990) Kadıköy District, Istanbul, Turkey (1996)	Szczecin (*Stettin*), Poland (1996) Oborishte District, Sofia, Bulgaria (2006)

Borough of KaMubukwana, Maputo, Mozambique (1995) Białołęka District Warsaw, Poland (2000)	Kaliningrad Oblast, Russia (2001) Hajnówka County, Poland (2001)	Jurbarkas (*Georgenburg*), Lithuania (2003)

District 15 Budapest, Hungary (1991) Tychy (*Tichau*), Poland (1992)	Kastrychnitski District, Minsk, Belarus (1993) Halton, United Kingdom (1994)	Lauingen (**BA**), Germany (1999)

Higashiōsaka, Japan (1960) Lahn-Dill District (**HE**), Germany (1961) Hamm (**NRW**), Germany (1962) Kassel (**HE**), Germany (1977) Holon, Israel (1980) Bottrop (**NRW**), Germany (1983)	Frogn municipality, Norway (1990) Schwalm-Eder District (**HE**), Germany (1992) Petrogradsky District, St. Petersburg, Russia (1994) Shinjuku, Japan (1994) Tsuwano, Japan (1995) Tourcoing, France (1995)	Fethiye, Turkey (1997) Chaoyang District, Beijing, China (2004) District 6 Budapest (Terézváros)/Terézváros, Hungary (2005) Central Administrative Okrug, Moscow, Russia (2006) Beyoğlu District, Istanbul, Turkey (2008)

Anderlecht, Belgium (1955) Boulogne-Billancourt, France (1955) London Borough of Hammersmith and Fulham, England (1955) Zaanstad, Netherlands (1955) Wetzlar (**HE**), Germany (1959)	Cologne (*Köln*) (**NRW**), Germany (1967) Leonberg (**BW**), Germany (1970) Bat-Yam, Israel (1978) Marino, Italy (1980) Ústí nad Orlicí, Czech Republic (1989)	Petrogradsky District, St. Petersburg, Russia (1991) 5th district of Prague, Czech Republic (2005) Borough of Cigli, Izmir, Turkey (2005))

Ashkelon, Israel (1994)	Kolobrzeg (*Kolberg*), Poland (1994)	Jalta, Ukraine (1992, resting)

Vogelsberg District (**HE**), Germany (1964) Greenwich, London, England (1966) Lauterbach (**HE**), Germany (1966) Antony, France (1966) Kiryat Ata, Israel (1976) Bad Steben (**BA**), Germany (1988) Melle (**LS**), Germany (1988) Blomberg (**NRW**), Germany (1990) Wroclaw (*Breslau*), Poland	Templewo, Poland Sulecin (*Zielenzig*), Poland Międzyrzecz (*Meseritz*), Poland Burkina Faso, the country Denia, Spain Province of Catalonia, Spain Minsk, Belarus Kiev, Ukraine Woltschja Gora, Belarus	Volgograd, Russia Sankt Petersburg, Russia Washington, D.C., USA Orkney Islands, Scotland Zeltingen-Rachtig (**RP**), Germany Lichtenfels District (**BA**), Germany Oberhavel District (**BB**), Germany Schmalkalden-Meiningen District (**TH**), Germany

Siegen-Wittgenstein District (**NRW**), Germany (1952) Siegen (**NRW**), Germany (1952) Luton, England (1959)	Asnières-sur-Seine, France (1959) Ashdod, Israel (1968)	Iznik, Turkey (1987) Nauen (**BB**), Germany (1988, then in the GDR)

Bad Godesberg District (**NRW**), Bonn, Germany (1962) Göttingen District (**LS**), Germany (1962) Hannoversch Münden (**LS**), Germany (1962) Rendsburg-Eckernförde District (**SH**), Germany (1964) Bremerhaven (**BN**), Germany (1965) Kiriat Bialik, Israel (1966 by S.) Nentershausen (**HE**), Germany (1966)	Hagen (**NRW**), Germany (1967) Brøndby, Denmark (1968) Cassino, Italy (1969) 12th arrondissement of Paris, France (1970) Westerwald District (**RP**), Germany (1970) Sderot, Israel (1975 by Z.) Ronneby, Sweden (1976)	Lüchow-Dannenberg District (**LS**), Germany (1979) Königs Wusterhausen (**BB**), Germany (1988 by Z.) Szilvásvárad in Heves County, Hungary (1989) Ordzhonikidzevsky Region, Kharkiv, Ukraine (1990) Kazimierz Dolny, Poland (1993) Sochos, Greece (1993) District 14 Budapest (Zugló), Hungary (2008)

Ahlen (**NRW**), Germany (1964) Bad Kreuznach District (**RP**), Germany (1964) Penzberg (**BA**), Germany (1964) Wuppertal (**NRW**), Germany (1964 by S.) London Borough of Barnet, United Kingdom (1955)	Koszalin (*Köslin*), Poland (1995) Nahariya, Israel (1970) Paderborn District (**NRW**), (1975) Teltow-Fläming District (**BB**), Germany (1991 by T.) Werra-Meißner District (**HE**), Germany (1957 by T., renewed 2001)	Amstelveen, Netherlands (2005) Borough of Mezitli, Mersin, Turkey (2012) Charenton-le-Pont, France Levallois-Perret, France (by S.)

East Norriton Township, USA (1991, administrative, by T.) Borough of Mokotów, Warszawa, Poland (1993, administrative, 2000 resting) Odernheim (**RP**), Germany (1997) Albinea, Italy (1997/1998)	Cajamarca, Peru (1998) Veszprém County, Hungary (2002) Subotica, Serbia (2002) Izola, Slovenia (2002)	Mürzzuschlag, Austria (2002) Olomouc (*Olmütz*), Czech Republic (2002) Olomouc Region, Czech Republic (2002) Innenstadt District (*Köln-Innenstadt*), Cologne (*Köln*), Germany (2005)

- Berlin-Zehlendorf, Berlin - 2001 amalgamated into the Berlin borough of Steglitz-Zehlendorf
- Bernburg, Saxony-Anhalt:

Fourmies, France (1967) Tarnowskie Góry, Poland (1968)	Rheine (**NRW**), Germany (1990) Anderson, Indiana, USA (1994)	Chomutov, Czech Republic Trakai, Lithuania

- Bernkastel-Kues, Rhineland-Palatinate:

Otmuchów, Poland (1993) Karlovy Vary, Czech Republic (2000)	Stadt Wehlen (**SX**), Germany	Milton Keynes, United Kingdom

- Biberach an der Riß, Baden-Württemberg:

Valence, France (1967) Tendring, United Kingdom	Asti, Italy Guernsey, United Kingdom	Telavi, Georgia (1987) Świdnica, Poland (1990)

- Bietigheim-Bissingen, Baden-Württemberg:

Kusatsu, Japan (1962) Sucy-en-Brie, France (1967)	Surrey Heath, United Kingdom (1971) Szekszárd, Hungary (1989)	Overland Park, Kansas, USA (1999)

- Bitburg, Rhineland-Palatinate:

Shelbyville, Kentucky, USA (1962) Diekirch, Luxembourg (1962)	Arlon, Belgium (1965) Rethel, France (1965)	Bad Köstritz (**TH**), Germany (1992)

- Bitterfeld, Saxony-Anhalt:

Vierzon, France (1959)	Dzerzhinsk, Russia (1996)	Kamienna Góra, Poland (2006)

- Blankenese, Hamburg:

  Gladsaxe, Denmark (1962)
- Blieskastel, Rhineland-Palatinate:

Le Creusot, France (1989)	Castellabate, Italy (2008)

- Bochum, North Rhine-Westphalia:

Sheffield, United Kingdom (1950)	Donetsk, Ukraine (1987)	Xuzhou, China (1994)
Oviedo, Spain (1980)	Nordhausen (**TH**), Germany (1990)	

- Böblingen, Baden-Württemberg:

Pontoise, France (1956)	Glenrothes, United Kingdom (1971)	Alba, Italy (1985)
Sittard-Geleen, Netherlands (1962)	Krems an der Donau, Austria (1972)	Sömmerda (**TH**), Germany (1988)
Bergama, Turkey, (1967)		

• Bonn, North Rhine-Westphalia:	La Paz, Bolivia (1996)	Chengdu, China (2009)
Oxford, United Kingdom (1947)	Opole, Poland (1997)	Salzburg, Austria
Tel Aviv-Jaffa, Israel (1983)	Cape Coast, Ghana	Ulan-Bator, Mongolia
Potsdam (**BB**), Germany (1988)	Aveiro, Portugal	
Budafok, Hungary (1991)	Yogyakarta, Indonesia	
Minsk, Belarus (1993)	Buchara, Uzbekistan	

twinned with Bad Godesberg:	Windsor and Maidenhead, United Kingdom (1960)
Saint-Cloud, France (1957)	Kortrijk, Belgium (1964)
Frascati, Italy (1960)	Yalova, Turkey (1969)

twinned with Hardtberg:	Villemomble, Paris, France (1967)

twinned with Beuel:	Mirecourt, France (1969)

- Bottrop, North Rhine-Westphalia:

  Blackpool, United Kingdom (2000)

- Bramsche, Lower Saxony:

Todmorden, United Kingdom	Biskupiec, Poland	Harfleur, France
Ra'anana, Israel (1978)		

- Braunfels, Hesse:

New Braunfels, Texas, USA	Rohrmoos-Untertal, Austria (1961)	Kiskunfélegyháza, Hungary (1992)
Bagnols-sur-Cèze, France (1959)	Newbury, United Kingdom (1963)	Feltre, Italy (1999)

- Braunschweig, Lower Saxony:

Bandung, Indonesia (1960)	Bath, United Kingdom (1971)	Kiryat Tiv'on, Israel (1985)
Nîmes, France (1962)	Sousse, Tunisia (1980)	Magdeburg (**SA**), Germany (1987)

- Bremen, Bremen:

Gdańsk (*Danzig*), Poland (1976)	Dalian, China (1985)	Haifa, Israel (1988)
Riga, Latvia (1985)	Rostock (**MV**), Germany (1987)	İzmir, Turkey (1995)

- Bremerhaven, Bremen:

Cherbourg-Octeville, France (1960) Grimsby, United Kingdom (1963)	Pori, Finland (1969) Frederikshavn, Denmark (1979)	Szczecin (*Stettin*), Poland (1990) Kaliningrad, Russia (1992)

- Brilon, North Rhine-Westphalia:

Hesdin, France (1965) Heusden-Zolder, Belgium (1971)	Thurso, United Kingdom (1980)	Buckow (**BB**), Germany (2003)

- Bückeburg, Lower Saxony:

Sablé-sur-Sarthe, France (1966)	Nieuwerkerk aan den IJssel, Netherlands (1974)

- Burghausen, Bavaria:

Fumel, France (1975) Ptuj, Slovenia (2001)	Hohenstein-Ernstthal (**SX**), Germany (2002) Sulmona, Italy (2006)	Schwechat, Austria Minakuchi, Japan

- Burgkirchen, Bavaria:

Wingen-sur-Moder, France (1977)	Kazincbarcika, Hungary (1998)	Sânnicolau Mare, Romania (2004)

- Butzbach, Hesse:

Eilenburg (**SX**), Germany (1990) Saint-Cyr-l'École, France (2008)	Collecchio, Italy	Teplá, Czech Republic

- Buxtehude, Lower Saxony:

Blagnac, France (1985)	Ribnitz-Damgarten (**MV**), Germany (1990)

C

- Calw, Baden-Württemberg:

  Latsch, Italy (1957)

- Castrop-Rauxel, North Rhine-Westphalia:

Wakefield, United Kingdom (1949) Delft, Netherlands (1950–2000)	Vincennes, France (1961) Kuopio, Finland (1965)	Zehdenik an der Havel (**BB**), Germany (1990) Nowa Ruda, Poland (1991)

- Celle, Lower Saxony:

Meudon, France (1953) Hämeenlinna, Finland (1971) Sumy, Ukraine (1990) Kwidzyn, Poland ("1990")	Quedlinburg (**SA**), Germany (1991) Tyumen, Russia (1994) Celle Ligure, Italy (2001) Mazkeret Batya, Israel (2008)	Holbæk, Denmark Tavistock, United Kingdom Tulsa, Oklahoma, USA

- Cham, Bavaria:

Cham, Switzerland (1981)	Klatovy, Czech Republic (1993)

- Chemnitz, Saxony:

Tampere, Finland (1961) Ljubljana, Slovenia (1966) Arras, France (1967) Timbuktu, Mali (1968)	Ústí nad Labem, Czech Republic (1970) Mulhouse, France (1981) Łódź, Poland (1981) Manchester, United Kingdom (1983)	Volgograd, Russia (1988) Düsseldorf (**NRW**), Germany (1988) Akron, Ohio, USA (1997) Taiyuan, China (1999)

- Coburg, Bavaria:

Garden City, New York, USA (1951) Oudenaarde, Belgium (1972)	Niort, France (1974) Gais, Italy (1977)	Isle of Wight, United Kingdom (1983) Cobourg, Ontario, Canada (1997)

- Coesfeld, North Rhine-Westphalia:
  De Bilt, Netherlands (1977)
- Cologne, North Rhine-Westphalia:

Liverpool, United Kingdom (1952) Esch-sur-Alzette, Luxembourg (1958) Liège (*Lüttich*), Belgium (1958) Lille, France (1958) Rotterdam, Netherlands (1958) Turin, Italy (1958) Kyoto, Japan (1963)	Tunis, Tunisia (1964) Turku, Finland (1967) Cluj-Napoca, Romania (1976) Tel Aviv, Israel (1979) Barcelona, Spain (1984) Beijing, China (1987) Corinto, Nicaragua (1988)	Cork, Ireland (1988) Indianapolis, Indiana, USA (1988) Thessaloniki, Greece (1988) Volgograd, Russia (1988) Katowice, Poland (1991) Bethlehem, Palestine (1996) Istanbul, Turkey (1997)

- Coswig, Saxony:

Ravensburg (**BW**), Germany (1990)	Lovosice nad Labem, Czech Republic (1995)

- Coswig an der Elbe, Saxony-Anhalt:
  Stadtallendorf (**HE**), Germany (1993)
- Cottbus, Brandenburg:

Montreuil-sous-Bois, France (1959) Grosseto, Italy (1967) Lipetsk, Russia (1974)	Zielona Góra, Poland (1975) Targovishte, Bulgaria (1975) Košice (*Kaschau*), Slovakia (1978)	Saarbrücken (**SL**), Germany (1987) Gelsenkirchen (**NRW**), Germany (1995) Nuneaton and Bedworth, United Kingdom (1999)

- Crailsheim, Baden-Württemberg:

Worthington, Minnesota, USA (first US-German town twinning, 1947) Pamiers, France (1969)	Jubarkas, Lithuania (2000)	Biłgoraj, Poland (2000)

- Crimmitschau, Saxony:

Bystřice nad Pernštejnem, Czech Republic (1971)	Wiehl (**NRW**), Germany (1990)

- Cuxhaven, Lower Saxony:

Vannes, France (1963) Penzance, United Kingdom (1967-1974, 2009) Penwith, United Kingdom (1974-2009) Nuuk, Greenland (1987-2011)	Piła, Poland (1996) Murmansk, Russia (2004) Hafnarfjörður, Iceland Ílhavo, Portugal	Vilanova de Arousa, Spain Binz (**MV**), Germany Sassnitz (**MV**), Germany

D

- Dachau, Bavaria:

Klagenfurt, Austria (1974) Fondi, Italy Bolzano, Italy (1990)	Tervuren, Belgium (2005) Renkum, Netherlands (2005) Areguá, Paraguay	Rosh HaAyin, Israel (planned)

- Darmstadt, Hesse:

Alkmaar, Netherlands (1958) Troyes, France (1958) Chesterfield, United Kingdom (1959) Graz, Austria (1968) Trondheim, Norway (1968)	Bursa, Turkey (1971) Płock, Poland (1988) Szeged, Hungary (1990) Gyönk, Hungary (1990) Freiberg (**SX**), Germany (1990)	Brescia, Italy (1991) Uzhhorod, Ukraine (1992) Liepāja, Latvia (1993) Saanen, Switzerland (1993) Logroño, Spain (2002)

- Datteln, Lower Saxony:

Cannock, United Kingdom (1971)	Genthin (**SA**), Germany (1990)

- Deggendorf, Bavaria:

  Neusiedl am See, Austria (1978)

- Delmenhorst, Lower Saxony:

  Toledo, Ohio, USA (2002)

- Dessau-Roßlau, Saxony-Anhalt:

Argenteuil, France (1959) Klagenfurt, Austria (1971) Ludwigshafen (**RP**), Germany (1988)	Ibbenbüren (**NRW**), Germany (1990) Gliwice, Poland (1992)	Nemenčinė, Lithuania (1995) Roudnice nad Labem, Czech Republic (2004)

- Detmold, North Rhine-Westphalia:

Saint-Omer, France (1969) Hasselt, Belgium (1976)	Zeitz (**SA**), Germany (1990) Savonlinna, Finland (2004)

- Dillenburg, Hesse:

  Diest, Belgium (1963)

- Dinkelsbühl, Bavaria:

Edenkoben (**RP**), Germany Porvoo, Finland	Guérande, France (1963) Sighişoara, Romania (2006)	Jingjiang, China Schmalkalden (**TH**), Germany (1990)

- Dinslaken, North Rhine-Westphalia:

Agen, France (1963)	Arad, Israel (1989)

- Döbeln, Saxony:

Givors, France Unna (**NRW**), Germany (1989)	Heidenheim (**BW**), Germany (1991)	Vyškov, Czech Republic (1992)

- Dormagen, North Rhine-Westphalia:

Toro, Spain (1994) Saint-André-lez-Lille, France (1973)	Kiryat Ono, Israel (1995)	Abilene, Texas, USA (must be wrong)

- Dorsten, North Rhine-Westphalia:

Crawley, United Kingdom (1973) Dormans, France (1981) Waslala, Nicaragua (1984)	Ernée, France (1985) Newtownabbey, United Kingdom (1988) Hainichen (**SX**), Germany (1990)	Hod HaSharon, Israel (1994) Rybnik, Poland (1994)

- Dortmund, North Rhine-Westphalia:

Amiens, France (1960) Leeds, United Kingdom (1969) Rostov, Russia (1973)	Buffalo, New York, USA (1977) Netanya, Israel (1980) Novi Sad, Serbia (1981)	Zwickau (**SX**), Germany (1988) Xian, China (1989)

- Dresden, Saxony:

Coventry, United Kingdom (1959) Saint Petersburg, Russia (1961) Wrocław, Poland (1963) Skopje, Macedonia (1967)	Florence, Italy (1971) Ostrava, Czech Republic (1971) Brazzaville, Republic of the Congo (1975) Hamburg (**HB**), Germany (1987)	Rotterdam, Netherlands (1988) Strasbourg, France (1990) Salzburg, Austria (1991) Columbus, Ohio, USA (1992)

- Duisburg, North Rhine-Westphalia:[3] [4]

Portsmouth, United Kingdom (1950)[3] [4] [5] [6] [7] Calais, France (1964)[3] [4] Fort Lauderdale, Florida, USA (2011) Lomé, Togo (1973)[3]	Wuhan, China (1982)[3] Vilnius, Lithuania (1985)[3] [4] Gaziantep, Turkey (2005)[3] [4]	Perm, Russia (2007)[4] Greenock, United Kingdom San Pedro Sula, Honduras (2008)

- Düren, North Rhine-Westphalia:

Valenciennes, France (1959) Cormeilles, France (1970) Altmünster, Austria (1971)	Gradačac, Bosnia and Hercegovina (2001) Stryj, Ukraine (2001)	Jinhua, China (2002) Ereğli, Turkey (2009)

- Düsseldorf, North Rhine-Westphalia:

Reading, United Kingdom (1975) Chemnitz (**SX**), Germany (1988)	Haifa, Israel (1988) Warsaw, Poland (1989)	Moscow, Russia (1992) Chongqing, China (2004)

E

- Eckernförde, Schleswig-Holstein:

Macclesfield, United Kingdom (1953) Hässleholm, Sweden (1958)	Tanga, Tanzania (1963) Nakskov, Denmark (1969)	Brzeg, Poland (1989) Bützow (**MV**), Germany (1990)

- Eggenfelden, Bavaria:

Carcassonne, France (1973)	Balatonalmádi, Hungary (2000)

- Eilenburg, Saxony:

Jihlava, Czech Republic (1987) Butzbach (**HE**), Germany (1990)	Anjalankoski, Finland (2000-2008) Tiraspol, Moldova (2002)	Rawicz, Poland (2001)

- Eisenhüttenstadt, Brandenburg:

  Saarlouis (**SL**), Germany (1982)

- Ellerhoop, Schleswig-Holstein:

  Hurst Green, United Kingdom [8] (1984)

- Eltville am Rhein, Hesse:

Montrichard, France (1965)	Passignano sul Trasimeno, Italy

twinned with district Hattenheim:	Arzens, France

- Elzach, Baden-Württemberg:

Telfs, Austria	Worthing, United Kingdom	Demre, Turkey

- Emden, Lower Saxony:

Arkhangelsk, Russia (1989)	London Borough of Hillingdon, United Kingdom (1961)	Haugesund, Norway (2009)

- Endingen am Kaiserstuhl, Baden-Württemberg:

Erstein, France (1970)	Coulonges-sur-l'Autize, France (1982)	Colonia Tovar, Venezuela

- Eppingen, Baden-Württemberg:

Wassy, France (1967)	Epping, United Kingdom (1981)	Szigetvár, Hungary (1992)

- Erfurt, Thuringia:

Győr, Hungary (1971) Vilnius, Lithuania (1972) Kalisz, Poland (1982) Mainz (**RP**), Germany (1988)	Lille, France (1991) Shawnee, Kansas, USA (1993) San Miguel de Tucumán, Argentina (1993) Lovech, Bulgaria (1996)	Haifa, Israel (2000) Xuzhou, China (2005)

- Erkelenz, North Rhine-Westphalia:

  Saint-James, France (1974)

- Erlangen, Bavaria:

Eskilstuna, Sweden (1961) Rennes, France (1964) Vladimir, Russia (1983/87)	Jena (**TH**), Germany (1987) Stoke-on-Trent, United Kingdom (1989) San Carlos, Nicaragua (1989)	Beşiktaş, district of Istanbul, Turkey (2003)

- Essen, North Rhine-Westphalia:

Sunderland, United Kingdom (1949) Tampere, Finland (1960)	Grenoble, France (1974) Nizhny Novgorod, Russia (1991)	Tel Aviv, Israel (1991)

- Esslingen am Neckar, Baden-Württemberg:

Neath Port Talbot, United Kingdom (1958) Vienne-sur-Isère, France (1958) Udine,Italy (1959) Norrköping, Sweden (1964) Coimbatore, India	Schiedam, Netherlands (1964) Sheboygan, Wisconsin, USA (1967) Velenje, Slovenia (1970) Molodechno, Belarus (1987)	Eger, Hungary (1991) Piotrków Trybunalski, Poland (1992)

- Euskirchen, North Rhine-Westphalia:

Charleville-Mézières, France (1964)	Basingstoke and Deane, United Kingdom (1986)	Kutztown, Pennsylvania, USA (2004)

- Eutin, Schleswig-Holstein:

Lawrence, Kansas, USA (1989)	Nykøbing Falster, Denmark	Putbus (**MV**), Germany (1990)

F

- Fellbach, Baden-Württemberg:

Tain-l'Hermitage, France (1964) Tournon-sur-Rhône, France (1973)	Erba, Italy (1978) Pécs, Hungary (1986)	Meißen (**SX**), Germany (1987)

- Feuchtwangen, Bavaria:

Lana, Italy	Morhange, France (1974)

- Flensburg, Schleswig-Holstein:

Carlisle, United Kingdom (1961) Neubrandenburg (**MV**), Germany (1987)	Słupsk, Poland (1988)	Tel Aviv, Israel (must be wrong)

- Forchheim, Bavaria:

Le Perreux-sur-Marne, France (1974) Roppen, Austria (1987)	Rovereto, Italy (1989) Pößneck (**TH**), Germany (1990)	Gherla, Romania (2000) Broumov, Czech Republic (2002)

- Frankfurt am Main, Hesse:

Lyon, France (1960)	Guangzhou, China (1988)	Budapest, Hungary (1990)
Birmingham, United Kingdom (1966)	Toronto, Ontario, Canada (1989)	Granada, Nicaragua (1991)
Milan, Italy (1970)	Leipzig (**SX**), Germany (1990)	Kraków, Poland (1991)
Tel Aviv, Israel (1980)	Prague, Czech Republic (1990)	

twinned with district Nieder-Eschbach:	Deuil-la-Barre, France (1967)

- Frankfurt an der Oder, Brandenburg:

Gorzów Wielkopolski, Poland (1975)	Vantaa, Finland (1987)	Yuma, Arizona, USA (1997)
Słubice, Poland (1975)	Heilbronn (**BW**), Germany (1988)	Kadima, Israel (1997)
Nîmes, France (1976)	Vitebsk, Belarus (1991)	

- Freiberg, Saxony:

Amberg (**BA**), Germany (1939)	Darmstadt (**HE**), Germany (1990)	Clausthal-Zellerfeld (**LS**), Germany
Gentilly, France (1960)	Wałbrzych, Poland (1991)	Nes Tzijona, Israel
Delft, Netherlands (1986)	Příbram, Czech Republic (1999)	

- Freiburg im Breisgau, Baden-Württemberg:

Besançon, France (1959)	Madison, Wisconsin, USA (1987)	Isfahan, Iran (2000)
Innsbruck, Austria (1963)	Matsuyama, Ehime, Japan (1989)	Tel Aviv, Israel
Padova, Italy (1967)	Lviv, Ukraine (1989)	
Guildford, United Kingdom (1979)	Granada, Spain (1991)	

- Freising, Bavaria:

Obervellach, Austria (1963)	Maria Wörth, Austria (1978)	Arpajon, France (1991)
Innichen, Italy (1969)	Waidhofen an der Ybbs, Austria (1986)	Škofja Loka (*Bischoflack*), Slovenia (2004)

- Friedberg, Hesse:

Bishop's Stortford, United Kingdom (1966-2012)	Villiers-sur-Marne, France (1966)	Magreglio, Italy

- Friedrichshafen, Baden-Württemberg:

Sarajevo, Bosnia and Herzegovina (1972)	Peoria, Illinois, USA (1976)	Delitzsch (**SX**), Germany (1990)
Saint-Dié-des-Vosges, France (1973)	Polotsk, Belarus (1990)	

- Fulda, Hesse:

Como, Italy (1960)	Sergiyev Posad, Russia (1991)	Litomerice, Czech Republic (2001)
Arles, France (1964)	Wilmington, Delaware, USA (1997)	

- Fürstenfeldbruck, Bavaria:

Livry Gargan, France (1967)	Wichita Falls, Texas, USA (1985)	Almuñécar, Spain (2005)
Cerveteri, Italy (1973)	Zadar, Croatia (1989)	

- Furth im Wald, Bavaria:

Furth bei Göttweig, Austria	Ludres, France	Domažlice, Czech Republic

- Fürth, Bavaria:

Renfrew, United Kingdom (1969) Bijeljina, Bosnia and Herzegovina (1987)	Limoges, France (1992) Marmaris, Turkey (1995)	Xylokastro, Greece (2005)

- Füssen, Bavaria:

Palestrina, Italy (1972) Helen, Georgia, USA (1978)	Bardu, Norway (1997) Numata, Japan (1998)	Airdrie, United Kingdom

G

- Gaggenau, Baden-Württemberg:

Annemasse, France (1970)	Sieradz, Poland (2000)

- Garbsen, Lower Saxony:

Farmers Branch, Texas, USA Hérouville-Saint-Clair, France (1966)	Rødding, Denmark Bassetlaw, United Kingdom	Schönebeck (Elbe) (**SA**), Germany (1990) Września, Poland

- Geesthacht, Schleswig-Holstein:

Hoogezand-Sappemeer, Netherlands (1966) Oldham, United Kingdom (1966–2004)	Plaisir, France (1975)	Kuldīga, Latvia (1991)

- Geislingen an der Steige, Baden-Württemberg:

Bischofswerda (**SX**), Germany (1990)	Montceau-les-Mines, France (1993)

- Georgsmarienhütte, Lower Saxony:

  Ramat HaSharon, Israel (2005)

- Gera, Thuringia:

Saint-Denis, France (1950) Sliven, Bulgaria (1965) Skierniewice, Poland (1965) Pskov, Russia (1969, renewed 1996)	Pilsen, Czech Republic (1970) Kuopio, Finland (1972) Timişoara, Romania (1984) Arnhem, Netherlands (1987, renewed 1991)	Rostov, Russia (1987) Nuremberg (**BA**), Germany (1988, renewed 1997) Fort Wayne, Indiana, USA (1992) Goražde, Bosnia and Herzegovina (2002)

- Gießen, Hesse:

Versailles, France (until 1985) Gödöllő, Hungary (1988) Hradec Králové, Czech Republic (1990)	Ferrara, Italy (2002) Kerkrade, Netherlands Netanya, Israel	San Juan del Sur, Nicaragua Waterloo, Iowa, USA Winchester, United Kingdom

  Wenzhou, China

- Goslar, Lower Saxony:

Arcachon, France (1965) Windsor and Maidenhead, United Kingdom (1969)	Beroun, Czech Republic (1989)	Brzeg, Poland (2000)

- Göttingen, Lower Saxony:

Cheltenham, United Kingdom (1951) Toruń, Poland (1978)	Pau, France (1982)	Lutherstadt Wittenberg (**SA**), Germany (1988)

- Grafenau, Bavaria:

Schärding, Austria (1976)	Kašperské Hory, Czech Republic (1991)

- Greifswald, Mecklenburg-Vorpommern:

Kotka, Finland (1959) Osnabrück (**LS**), Germany (1988)	Lund, Sweden (1990) Hamar, Norway (1997)	Newport News, Virginia, USA (2007)

- Gronau, North Rhine-Westphalia:

Epe, Netherlands	Bromsgrove, United Kingdom (1980)	Mezőberény, Hungary (1987)

- Groß-Gerau, Hesse:

Brignoles, France (1959) Bruneck (Brunico), Italy (1959)	Tielt, Belgium (1959)	Szamotuły, Poland (2000)

- Großostheim, Bavaria:

Olympia, Greece (2001) Carbon-Blanc, France	Saulxures-sur-Moselotte, France	Hamoir, Belgium

- Grünberg, Hesse:

Condom, France (1973)	Mrągowo, Poland (1993)

- Grünstadt, Rhineland-Palatinate:

Bonita Springs, Florida, USA Carrières-sur-Seine, France (1974)	Hermsdorf (**TH**), Germany Greenville, Ohio, USA	Westerburg (**RP**), Germany Peine (**LS**), Germany (2003)

- Gütersloh, North Rhine-Westphalia:

Châteauroux, France (1977) Broxtowe, United Kingdom (1978)	Grudziądz, Poland (1989) Falun, Sweden (1994)	Rzhev, Russia (2008)

- Gummersbach, North Rhine-Westphalia:

La Roche-sur-Yon, France (1968)	Burg (**SA**), Germany (1990)	Afandou, Rhodes, Greece (2001)

H

- Hagen, North Rhine-Westphalia:

Liévin, France (1960) Kouvola, Finland (1963–2009) Montluçon, France (1965)	borough of Steglitz-Zehlendorf, Berlin (**BE**), Germany (1967) Bruck an der Mur, Austria (1975) Smolensk, Russia (1985)	Modi'in, Israel (1997)

- Halberstadt, Saxony-Anhalt:

Wolfsburg (**LS**), Germany (1989) Banská Bystrica, Slovakia (1998)	Náchod, Czech Republic (1998)	Villars, France (2003)

- Halle an der Saale, Saxony-Anhalt:

Oulu, Finland (1968) Coimbra, Portugal (1974) Linz, Austria (1975)	Grenoble, France (1976) Karlsruhe (**BW**), Germany (1987) Hildesheim (**LS**), Germany (1992)	Ufa, Republic of Bashkortostan, Russia (1997) Jiaxing, China (2009) Savannah, Georgia, USA (2011)

- Hameln, Lower Saxony:

Saint-Maur-des-Fossés, France (1968) Torbay, United Kingdom (1973)	Quedlinburg (**SA**), Germany (1991)	Kalwaria Zebrzydowska, Poland (2001)

- Hamburg, Hamburg:

Saint Petersburg, Russia (1957) Marseille, France (1958) Shanghai, China (1986)	Dresden (**SX**), Germany (1987) Osaka, Japan (1989) León, Nicaragua (1989)	Prague, Czech Republic (1990) Chicago, Illinois, USA (1994) Dar es Salaam, Tanzania (2010)

- Hanover, Lower Saxony:

Hanover Park, Illinois, USA Bristol, United Kingdom (1947)[9] Perpignan, France (1960)	Rouen, France (1966) Blantyre, Malawi (1968) Poznań, Poland (1979)[10]	Hiroshima, Japan (1983)[11] [12] Leipzig (**SX**), Germany (1987)[13]

- Hechingen, Baden-Württemberg:[14]

Joué-lès-Tours, France (1973)	Limbach-Oberfrohna (**SX**), Germany (1990)	Hódmezövásárhely, Hungary (1994)

- Heidelberg, Baden-Württemberg:

Montpellier, France (1961) Cambridge, United Kingdom (1965)	Rehovot, Israel (1983) Simferopol, Ukraine (1991)	Bautzen (**SX**), Germany (1991) Kumamoto, Japan (1992)

- Heidenheim, Baden-Württemberg:

Clichy, France (1958) Sankt Pölten, Austria (1968)	Newport, United Kingdom (1981) Sisak, Croatia (1988)	Döbeln (**SX**), Germany (1991) Jihlava, Czech Republic (2002)

- Heilbronn, Baden-Württemberg:

Béziers, France (1965) Neath Port Talbot, United Kingdom (1966)	Solothurn, Switzerland (1981) Stockport, United Kingdom (1982)	Słubice, Poland (1998) Frankfurt (Oder) (**BB**), Germany (1988)

- Helmstedt, Lower Saxony:

Vitré, France (1978) Chard, United Kingdom (1980) Albuquerque, New Mexico, USA (1983)	Fiuggi, Italy (1986) Haldensleben (**SA**), Germany (1990) Svetlahorsk, Belarus (1991)	Orăştie, Romania (2002)

- Hemsbach, Baden-Württemberg:

Bray-sur-Seine, France (1972)	Wareham, United Kingdom (1986)	Mücheln (**SA**), Germany (1990)

- Herford, North Rhine-Westphalia:

Hinckley, United Kingdom (1972) Fredericia, Denmark (1987)	Quedlinburg (**SA**), Germany (1990)	Gorzów Wielkopolski, Poland (1990)

- Herne, North Rhine-Westphalia:

Hénin-Beaumont, France (1954) Wakefield, United Kingdom (1956)	Ometepe, Nicaragua (1988) Belgorod, Russia (1990)	Eisleben (**SA**), Germany (1990) Konin, Poland (1991)

- Hersbruck, Bavaria:

  Lossiemouth, United Kingdom (1972)
- Heusenstamm, Hesse:

Saint-Savin, France (1969) Tonbridge, United Kingdom (1984)	Malle, Belgium (1991)	Ladispoli, Italy (2001)

- Hildesheim, Lower Saxony:

Angoulême, France (1965) El Minia, Egypt (1979) Weston-super-Mare, United Kingdom (1983)	Padang, Indonesia (administrative cooperation, 1988) Gelendzhik, Russia (1992)	North Somerset, United Kingdom (1997) Pavia, Italy (2000)

- Hockenheim, Baden-Württemberg:

Commercy, France (1970)	Hohenstein-Ernstthal (**SX**), Germany (1990)	Mooresville, North Carolina, USA (2002)

- Hohen Neuendorf, Brandenburg:

Fürstenau (**LS**), Germany (1991) Maing, France (1992)	Müllheim (**BW**), Germany (1992)	Janów Podlaski, Poland (1995)

- Holzminden, Lower Saxony:

  Leven, Fife, Scotland

- Holzwickede, North Rhine-Westphalia:

Louviers, France (1977/78)	Weymouth, United Kingdom (1986/87)

- Hückelhoven, North Rhine-Westphalia:
  - Hartlepool, United Kingdom (1973)
- Husum, Schleswig-Holstein:

Kidderminster, United Kingdom (1976) Heilbad Heiligenstadt (**TH**), Germany	Gentofte, Denmark	Trzcianka, Poland

I

- Ibbenbüren, North Rhine-Westphalia:

Gourdon, France Hellendoorn, Netherlands	Prievidza, Slovakia Dessau-Roßlau (**SA**), Germany (1990)	Jastrzębie Zdrój, Poland (2007)

- Illertissen, Baden-Württemberg:

Carnac, France	Loket, Czech Republic (1953)

- Ingolstadt, Bavaria:

Carrara, Italy (1962) Kirkcaldy, United Kingdom (1962) Grasse, France (1963)	Murska Sobota, Slovenia (1979) Moscow, Russia (1995) Manisa, Turkey (1998)	Kragujevac, Serbia (2003) Opole, Poland (2005) Györ, Hungary (2008)

- Iserlohn, North Rhine-Westphalia:

Almelo, Netherlands (1954) Biel, Switzerland (1959) Auchel, France (1966)	Hall in Tirol, Austria (1967) Nyíregyháza, Hungary (1989) Novocherkassk, Russia (1990)	Glauchau (**SX**), Germany (1991) Chorzów, Poland (1995)

- Isny, Bavaria:

Andrychów, Poland (1998) Flawil, Switzerland	Notre-Dame-de-Gravenchon, France Sotkamo, Finland	Street, Somerset, United Kingdom

J

- Jade, Lower Saxony:
  - Bakonynána, Hungary (2001)
- Jena, Thuringia:

Lugoj, Romania (1983, renewed 1993) Porto, Portugal (1984) Erlangen (**BA**), Germany (1987)	Berkeley, California, USA (1989) San Marcos, Nicaragua (1996) Aubervilliers, France (1999)	Kamëz, Albania[15] (2009) Beit Jala, Palestine (2011)

- Jever, Lower Saxony:

Zerbst (**SA**), Germany (1990)	Cullera, Spain (1998)

- Jülich, North Rhine-Westphalia:
  - Haubourdin, France (1964)
- Jüterbog, Brandenburg:

Aßlar (**HE**), Germany (1991)	Waldbröl (**NRW**), Germany

K

- Kahl am Main, Bavaria:

Villefontaine, France (1981)	Budakalász, Hungary (1998)

- Kaiserslautern, Rhineland-Palatinate:

Rotherham, United Kingdom Davenport, Iowa, USA[16] (1960) Douzy, France (1967) Saint-Quentin, Aisne, France (1967) London Borough of Newham, United Kingdom (1974)	Bunkyo-ku, district of Tokyo, Japan (1988) Brandenburg an der Havel (**BB**), Germany (1988) Pleven, Bulgaria (1999) Columbia, South Carolina, USA (2000) Silkeborg, Denmark (2000)	Guimarães, Portugal (2000) Banja Luka, Bosnia and Herzegovina (2003) Bitola, Macedonia Igualada, Spain

- Kamenz, Saxony:

Alzey (**RP**), Germany (1990)	Karpacz, Poland	Kolín, Czech Republic

- Kandern, Baden-Württemberg:
  - Soufflenheim, France (1985)
- Karlsruhe, Baden-Württemberg:

Nancy, France (1955) Nottingham, United Kingdom (1969)	Halle an der Saale (**SA**), Germany (1987) Timişoara, Romania (1997)	Krasnodar, Russia (1998)

- Kassel, Hesse:

Florence, Italy (1952) borough of Mitte, Berlin (**BE**), Germany (1962) Mulhouse, France (1965) Rovaniemi, Finland (1972)	Västerås, Sweden (1972) Yaroslavl, Russia (1988) Arnstadt (**TH**), Germany (1989) Ramat Gan, Israel (1990)	İzmit, Turkey (1999) Montana, Bulgaria (2007)

- Kaufbeuren, Bavaria:

Ferrara, Italy (1991)	Szombathely, Hungary (1992)

- Kehl, Baden-Württemberg:

Montmorency, France (1968)	Carmarthenshire, United Kingdom (1995)	Kotor, Montenegro (2000)

- Kelheim, Bavaria:

Ambarès-et-Lagrave, France (1989)	Soave, Italy (2006)

- Kempten, Bavaria:

Bad Dürkheim (**RP**), Germany (1951)	Sopron, Hungary (1987)	Sligo, Ireland (1990)
Quiberon, France (1971)	Trento, Italy (1987)	

- Kiel, Schleswig-Holstein:

Brest, France (1964)	Gdynia, Poland (1985)	Kaliningrad, Russia (1992)
Coventry, United Kingdom (1967)	Tallinn, Estonia (1986)	Sovetsk, Russia (1992)
Vaasa, Finland (1967)	Stralsund (**MV**), Germany (1987)	

- Kierspe

  Denton, Greater Manchester, United Kingdom (2012)

  Montigny-le-Bretonneux, France (1988)

- Kitzingen, Bavaria:

Montevarchi, Italy (1984)	Neuern, Czech Republic	Trzebnica, Poland (2009)
Prades, France (1984)		

- Koblenz, Rhineland-Palatinate:

Nevers, France (1963)	Maastricht, Netherlands (1981)	Petah Tikva, Israel (2000)
London Borough of Haringey, United Kingdom (1969)	Novara, Italy (1991)	Varaždin, Croatia (2007)
Norwich, United Kingdom (1978)	Austin, Texas, USA (1992)	

- Konstanz, Baden-Württemberg:

Fontainebleau, France (1960)	Lodi, Italy (1986)	Pančevo, Serbia
Richmond, London, United Kingdom (1983)	Colwyn, United Kingdom	
Tábor, Czech Republic (1984)	Suzhou, China	

- Korbach, Hesse:

Avranches, France (1963)	Pyrzyce, Poland	Vysoké Mýto, Czech Republic
Waltershausen (**TH**), Germany (1990)		

- Krefeld, North Rhine-Westphalia:

Venlo, Netherlands (1964)	Leiden, Netherlands (1974)	Ulyanovsk, Russia (1993)
Leicester, United Kingdom (1969)	Charlotte, North Carolina, USA (1986)	Kayseri, Turkey (2009)
Dunkirk, France (1974)	Beeskow (**BB**), Germany (1990)	

- Künzelsau, Baden-Württemberg:

  Marcali, Hungary (1992)

L

- Lahr, Baden-Württemberg:

Dole, France (1962)	Belleville, Ontario, Canada (1971)	Alajuela, Costa Rica (2006)

- Landau in der Pfalz, Rhineland-Palatinate:

Ribeauvillé (*Rappoldsweiler*), France (1960)	Haguenau, France (1963) Kigoma, Rwanda (1984)	Ruhango, Rwanda (2001)

- Landshut, Bavaria:

Elgin, United Kingdom (1956) Compiègne, France (1962)	Ried im Innkreis, Austria (1974) Schio, Italy (1981)	Sibiu (*Hermannstadt*), Romania (2002)

- Langen, Hesse:

Long Eaton, United Kingdom (1968) Romorantin-Lanthenay, France (1970)	Tarsus, Turkey (1991)	Aranda de Duero, Spain (2006)

- Langenhagen, Lower Saxony:

Rodewisch (**SX**), Germany (1989, renewed 1992) Głogów (*Glogau*), Poland (1993) Bijeljina, Bosnia and Herzegovina (2007)	Southwark, United Kingdom Novo Mesto, Slovenia Le Trait, France	Stadl-Paura, Austria Joinville, Brazil

- Laufen, Bavaria:

Brioude, France (1982)	Laufen, Switzerland	Leobendorf, Austria

- Leer, Lower Saxony:

Trowbridge, United Kingdom (1989)	Elbląg, Poland (1992)

- Leichlingen, North Rhine-Westphalia:
  - Funchal, Portugal
- Leipzig, Saxony:

Kiev, Ukraine (1961, renewed 1992) Bologna, Italy (1962, renewed 1997) Kraków, Poland (1973, renewed 1995) Brno, Czech Republic (1973, renewed 1999) Lyon, France (1981)	Thessaloniki, Greece (1984) Hanover (**LS**), Germany (1987) Nanjing, China (1988) Frankfurt am Main (**HE**), Germany (1990) Birmingham, United Kingdom (1992)	Houston, Texas, USA (1993) Travnik, Bosnia and Herzegovina (2003) Addis Ababa, Ethiopia (2004)

- Leutershausen, Bavaria:
  - Szendrö, Hungary (2000)
- Leutkirch, Baden-Württemberg:

Lamalou-les-Bains, France (1982) Hérépian, France (1982)	Bédarieux, France (1982)	Castiglione delle Stiviere, Italy (1995)

- Liederbach am Taunus, Hesse:

Villebon-sur-Yvette, France (1985) Frauenwald (**TH**), Germany (1990)	Verwood, United Kingdom (1992) Pietrowice Wielkie (*Groß Peterwitz*), Poland	Saldus, Latvia

- Lindlar, North Rhine-Westphalia:

Shaftesbury, United Kingdom (1981)	Brionne, France (1983)	Kaštela, Croatia (1987)

- Linz am Rhein, Baden-Württemberg:

Linz, Austria (1920)	Marietta, Georgia, USA (1965)	Pornic, France (1987)

- Lollar, Hesse:
  Brassac-les-Mines, France (1987)
- Lörrach, Baden-Württemberg:

Sens, France (1966) Senigallia, Italy (1986)	Meerane (**SX**), Germany (1990)	Chester, United Kingdom (2002)

- Lübbecke, North Rhine-Westphalia:

Bayeux, France (1968) Dorchester, United Kingdom (1973)	Tiszakécske, Hungary (1989)	Bad Liebenwerda (**BB**), Germany (1990)

- Lüdenscheid, North Rhine-Westphalia:
  Brighouse, United Kingdom (1950)
- Ludwigsburg, Baden-Württemberg:
  Montbéliard, France (1951)
- Ludwigsfelde, Brandenburg:
  Zdzieszowice, Poland (1998)
- Ludwigshafen, Rhineland-Palatinate:

Pasadena, California, USA (1958) Lorient, France (1963)	London Borough of Havering, United Kingdom (1971) Sumqayıt, Azerbaijan (1987)	Dessau (**SA**), Germany (1988) Antwerp, Belgium (1999)

- Ludwigslust, Mecklenburg-Vorpommern:

Ahrensburg (**SH**), Germany (1990)	Kamskoye Ustye, Russia (1994)	Muscatine, Iowa, USA (2004)

- Lüneburg, Lower Saxony:

Scunthorpe, United Kingdom (1960) Kulmbach (**BA**), Germany (1967) Naruto, Japan (1974)	Clamart, France (1975) Ivrea, Italy (1988) Köthen (**SA**), Germany	Viborg, Denmark (1992) Tartu, Estonia (1993)

- Lünen, North Rhine-Westphalia:

Zwolle, Netherlands (1963) Salford, United Kingdom (1966)	Kamień Pomorski (*Cammin*), Poland Panevėžys, Lithuania (1990)	Demmin (**MV**), Germany Bartın, Turkey

M

- Mainz, Rhineland-Palatinate:

Watford, United Kingdom (1956) Dijon, France (1957) Zagreb, Croatia (1967)	Valencia, Spain (1978) Baku, Azerbaijan (1984) Haifa, Israel (1987)	Erfurt (**TH**), Germany (1988) Louisville, Kentucky, USA (1994)

- Malchow, Mecklenburg-Vorpommern:

Quickborn (**SH**), Germany (1990) Moormerland (**LS**), Germany (1990)	Langballig (**SH**), Germany	Gniewino (*Gnewin*), Poland

- Mannheim, Baden-Württemberg:

Swansea, United Kingdom (1957) Bydgoszcz, Poland (1991)	Chişinău, Moldova (1989) Klaipėda, Lithuania, (2002)	Zhenjiang, China (2004) Windsor, Ontario, Canada

- Marbach am Neckar, Baden-Württemberg:

L'Isle-Adam, France (1987) Washington, Missouri, USA (1990)	Tongling, China	St. George, Utah, USA

- Marburg, Hesse:

Poitiers, France (1961) Maribor, Slovenia (1969)	Sfax, Tunisia (1971) Eisenach (**TH**), Germany (1988)	Northampton, United Kingdom (1992) Sibiu (*Hermannstadt*), Romania (2005)

- Markgröningen, Baden-Württemberg:
  - Saint-Martin-de-Crau, France (1989)
- Meerbusch, North Rhine-Westphalia:
  - Fouesnant, France (1970)
- Merzig, Saarland:

Saint-Médard-en-Jalles, France (1986)	Luckau (**BB**), Germany (2009)

- Mönchengladbach, North Rhine-Westphalia:

North Tyneside, United Kingdom (1958) Roubaix, France (1969)	Thurrock, United Kingdom (1969) Verviers, Belgium (1970)	Bradford, United Kingdom (1971)

- Mosbach, Baden-Württemberg:

Château-Thierry, France (1974) Pößneck (**TH**), Germany (1989)	Finike, Turkey (1994) Lymington, United Kingdom (1996)	Pesthidegkút, Hungary (1998)

- Memmingen, Bavaria:

Glendale, Arizona, USA (1976) Province of Teramo, Italy (1981) Teramo, Italy (1981)	Auch, France (1990) Eisleben (**SA**), Germany (1990) Kiryat Shmona, Israel (2009)	Karataş, Turkey (2009) Litzelsdorf, Austria (2009) Chernihiv, Ukraine (2009)

- Mühlhausen, Thuringia:

Tourcoing, France (1961) Eschwege (**HE**), Germany (1989)	Münster (**NRW**), Germany (1990) Kronstadt, Russia (1995)	Saxonburg, Pennsylvania, USA (2008)

- Mülheim, North Rhine-Westphalia:

Darlington, United Kingdom (1953) Tours, France (1962) Kuusankoski, Finland (1972-2008)	Opole, Poland (1989) Kfar Sava, Israel (1993) Istanbul-Beykoz, Turkey (2007)	Kouvola, Finland (2009) Qalqilyah, Palestine

- Munich, Bavaria:

Edinburgh, United Kingdom (1960) Verona, Italy (1960) Bordeaux, France (1964)	Sapporo, Japan (1972) Cincinnati, Ohio, USA (1989) Kiev, Ukraine (1989)	Harare, Zimbabwe (1996) Hyderabad, India (2005)

- Münster, North Rhine-Westphalia:

Braniewo, Poland (1954) York, United Kingdom (1958) Orléans, France (1960) Kristiansand, Norway (1967)	Monastir, Tunisia (1969) Rishon LeZion, Israel (1981) Fresno, California, USA (1986) Ryazan, Russia (1989)	Mühlhausen (**TH**), Germany (1990) Lublin, Poland (1991)

N

- Nauen, Brandenburg:
  borough of Spandau, Berlin (**BE**), Germany (1988)
- Naumburg, Saxony-Anhalt:
  Aachen (**NRW**), Germany (1988)
- Neubukow, Mecklenburg-Vorpommern:

Steinfurt (**NRW**), Germany (1990)	Reinfeld (**SH**), Germany (1991)

- Neustrelitz, Mecklenburg-Vorpommern:

Rovaniemi, Finland (1963) Szczecinek, Poland (1987)	Schwäbisch Hall (**BW**), Germany (1988)	Chaykovsky, Russia (1993)

- Neu Wulmstorf, Lower Saxony:
  Nyergesújfalu, Hungary (1991)
- Norderstedt, Schleswig-Holstein:

Maromme, France (1966) Oadby and Wigston, United Kingdom (1977)	Zwijndrecht, Netherlands (1981)	Kohtla-Järve, Estonia (1989)

- Nordhausen, Thuringia:

Charleville-Mézières, France Bochum (**NRW**), Germany (1990)	Beit Shemesh, Israel (1992)	Ostrów Wielkopolski, Poland (1995)

- Nördlingen, Bavaria:

  Wagga Wagga, Australia (1967)

- Nuremberg, Bavaria:

Nice, France (1954) Kraków, Poland (1979) Skopje, Macedonia (1982) Glasgow, United Kingdom (1985) San Carlos, Nicaragua (1985)	Gera (**TH**), Germany (1988) Kharkiv, Ukraine (1990) Prague, Czech Republic (1990) Hadera, Israel (1995) Antalya, Turkey (1997)	Shenzhen, China (1997) Kavala, Greece (1998) Atlanta, Georgia, USA (1998) Venice, Italy (1999) Córdoba, Spain (planned for 2010)

- Nürtingen, Baden-Württemberg:

Oullins, France (1962) Rhondda Cynon Taff, United Kingdom (1968)	Soroskár, Hungary (1988)	Zerbst (**SA**), Germany (1992)

O

- Oberursel, Hesse:

Rushmoor, United Kingdom (1989) Epinay-sur-Seine, France (1964)	Koggenland/Ursem, Netherlands (1971)	Lomonosov (*Oranienbaum*), Russia (2004)

- Ochsenfurt, Bavaria:

Coutances, France (1983) Wimborne, United Kingdom (1989)	Colditz (**SA**), Germany (1990) Ropczyce, Poland	Zábřeh, Czech Republic

- The Odenwald region, Hesse, Bavaria and Baden-Württemberg:

  Falkirk, United Kingdom

- Oestrich-Winkel, Hesse:

Tokaj, Hungary	Denicé, France

- Olfen, North Rhine-Westphalia:

  La Grange, Texas, USA

- Olbernhau, Saxony:

Bagnolo Mella, Italy Brie-Comte-Robert, France	Litvínov, Czech Republic	Stadtbergen (**BA**), Germany (1989)

- Osnabrück, Lower Saxony:

Haarlem, Netherlands (1961) Angers, France (1964) Derby, United Kingdom (1976) Greifswald (**MV**), Germany (1988)	Evansville, Indiana, USA (1991) Tver, Russia (1991) Gwangmyeong, South Korea (1997) Çanakkale, Turkey (2004)	Vila Real, Portugal (2005) Hefei, China (2006)

P

- Paderborn, North Rhine-Westphalia:

Le Mans, France (traditionally 836, officially 1967) Bolton, United Kingdom (1975) Belleville, Illinois, USA (1990)	Pamplona, Spain (1992) Przemyśl, Poland (1993) Debrecen, Hungary (1994)	Qingdao, China (2003) Ludwigsfelde (**BB**), Germany (informally)

- Passau, Bavaria:

Hackensack, New Jersey, USA (1952) Dumfries, United Kingdom (1957) Cagnes-sur-Mer, France (1973) Krems an der Donau, Austria (1974)	Akita, Japan (1984) Málaga, Spain (1987) České Budějovice, Czech Republic (1993) Liuzhou, China (1999)	Veszprém, Hungary (1999) Montecchio Maggiore, Italy (2003)

- Pfungstadt, Hesse:

Retford, United Kingdom (1979) Figline Valdarno, Italy (1993)	Gradignan, France (1996)	Hévíz, Hungary (2005)

- Pinneberg, Schleswig-Holstein:

Rockville, Maryland, USA (1957)	Nzega-District, Tanzania	Primorsk (*Fischhausen*), Russia

- Pirmasens, Rhineland-Palatinate:

  Poissy, France (1965)

- Pocking, Bavaria:

  Metulla, Israel

- Potsdam, Brandenburg:

Opole, Poland (1973) Bobigny, France (1974) Jyväskylä, Finland (1985)	Bonn (**NRW**), Germany (1988) Sioux Falls, South Dakota, USA (1990) Perugia, Italy (1990)	Luzern, Switzerland (2002)

- Preetz, Schleswig-Holstein:

Blandford, United Kingdom (1979) Stavenhagen (**MV**), Germany (1990)	Tapa, Estonia (1991)	Neman (*Ragnit*), Russia (1991)

- Pulheim, North Rhine-Westphalia:

Guidel, France (1969)	Fareham, United Kingdom (1984)

Q

- Quickborn, Schleswig-Holstein:

Boxholm, Sweden (1974)	Uckfield, United Kingdom (1990)	Malchow (**MV**), Germany (1990)

R

- Rastatt, Rastatt District, Baden-Württemberg:

Orange, France (1965) Fano, Italy (1985)	Guarapuava, Brazil (1987) Woking, United Kingdom (2001)	New Britain, Connecticut, USA (1984) Ostrov (*Schlackenwerth*), Czech Republic (1991)

- Ratingen, Mettmann District, North Rhine-Westphalia:

Maubeuge, France (1958) Le Quesnoy, France (1963) Blyth Valley, United Kingdom (1967)	Vermillion, South Dakota, USA (1969) Cramlington, United Kingdom (1974) Kokkola, Finland (1989)	Beelitz (**BB**), Germany (1990) Gagarin, Russia (1998) Wuxi, China (2007)

- Regensburg, Bavaria:

Aberdeen, United Kingdom (1955) Brixen, Italy (1959) Clermont-Ferrand, France (1969)	Tempe, Arizona, USA (1981) Odessa, Ukraine (1990) Plzeň, Czech Republic (1993)	Budavár (district of Budapest), Hungary (2005) Qingdao, China (2009)

- Remscheid, North Rhine-Westphalia:

Quimper, France (1971) Wansbeck, United Kingdom (1978)	Prešov, Slovakia (1989) Pirna (**SX**), Germany (1989)	Kirsehir, Turkey Schmalkalden (**TH**), Germany

- Rendsburg, Schleswig-Holstein:

Hereford, United Kingdom (1950)	Aalborg, Denmark (1967)	Haapsalu, Estonia (1989)

- Reutlingen, Baden-Württemberg:

Aarau, Switzerland (1986)	Szolnok, Hungary (1990)

- Rheinfelden, Baden-Württemberg:

Rheinfelden, Switzerland (informally) Fécamp, France (1963)	Vale of Glamorgan, United Kingdom (1968) Neumarkt, Italy (1968)	Mouscron, Belgium (1981) Ludwigsfelde (**BB**), Germany (informally)

- Rinteln, Lower Saxony:

Kendal, United Kingdom (1992)	Sławno (*Schlawe*), Poland

- Rockenhausen, Rhineland-Palatinate:

Rognac, France (1974)	Głubczyce (*Leobschütz*), Poland (2003)

- Rosenheim, Bavaria:

Briançon, France (1974) Lazise, Italy (1979)	Beersheva, Israel (1983) Greiz (**TH**), Germany (1991)	Ichikawa, Japan (2004)

- Rostock, Mecklenburg-Vorpommern:

Szczecin (*Stettin*), Poland (1957) Turku, Finland (1959) Dunkirk, France (1960) Riga, Latvia (1961) Antwerp, Belgium (1963)	Århus, Denmark (1964) Gothenburg, Sweden (1965) Bergen, Norway (1965) Varna, Bulgaria (1966) Rijeka, Croatia (1966)	Bremen (**BN**), Germany (1987) Dalian, China (1988) Raleigh, North Carolina, USA (2001)

- Rüdersdorf bei Berlin, Brandenburg:

  Pierrefitte-sur-Seine, France (1965)

- Rüdesheim, Hesse:

Meursault, France	Mezőkövesd, Hungary (1991)	Swanage, United Kingdom

twinned with Assmannshausen:	Juliénas, France

twinned with Presberg:	Oingt, France

S

- Salzgitter, Lower Saxony:

Imatra, Finland (1970) Swindon, United Kingdom (1975)	Créteil, France (1980) Stary Oskol, Russia (1987)	Gotha (**TH**), Germany (1988)

- Salzwedel, Saxony-Anhalt:

Wesel (**NRW**), Germany (1990)	San Vito dei Normanni, Italy (1990)	Felixstowe, United Kingdom (1994)

- Saarbrücken, Saarland:

Nantes, France (1965)	Tbilisi, Georgia (1975)	Cottbus (**BB**), Germany (1987)

- Saarlouis, Saarland:

Saint-Nazaire, France (1969)	Eisenhüttenstadt (**BB**), Germany (1986, first town partnership across the German-German border)

- Sankt Wendel, Saarland:

Rezé, France (1973)	São Vendelino, Brazil (2003)	Balbriggan, Ireland (2007)

- Schmallenberg, North Rhine-Westphalia:

Wimereux, France (1972)	Burgess Hill, England (1988)

- Schopfheim, Baden-Württemberg:

Poligny, France (1967) Ronneby, Sweden (1987)	Kleinmachnow (**BB**), Germany (1996)	Dikome, Cameroon (2000)

- Schriesheim, Baden-Württemberg:
  Uzès, France (1984)
- Schrobenhausen, Bavaria:
  Bridgnorth, United Kingdom (1992)
- Schweinfurt, Bavaria:

Motherwell, Scotland	Châteaudun, France	Seinäjoki, Finland

- Schwäbisch Gmünd, Baden-Württemberg:

Barnsley, United Kingdom (1971) Antibes, France (1976)	Bethlehem, Pennsylvania, USA (1991) Szekesfehervar, Hungary (1991)	Faenza, Italy (2001)

- Schwäbisch Hall, Baden-Württemberg:

Epinal, France (1964) Lappeenranta, Finland (1985)	Loughborough, United Kingdom (1966) Neustrelitz (**MV**), Germany (1988)	Zamość (*Zamosch*), Poland (1989)

- Seesen, Lower Saxony:

Wantage, United Kingdom (1978) Thale (**SA**), Germany (1990)	Carpentras, France (1993)	Montecorvino Rovella, Italy (2006)

- Seligenstadt, Hesse:

Triel-sur-Seine, France (1967) Heel, Netherlands (1972-2008)	Brookfield, Wisconsin, USA (2008)	Piedimonte Matese, Italy (2010)

- Sindelfingen, Baden-Württemberg:

Schaffhausen, Switzerland (1952) Corbeil-Essonnes, France (1961) Sondrio, Italy (1962)	Dronfield, United Kingdom (1971) Győr, Hungary (1987) Torgau (**SX**), Germany (1987)	Chełm, Poland (2001)

- Soest, North Rhine-Westphalia:

Guérard, France (1967) Bangor, Wales (1973) Kampen, Netherlands (1992) Herzberg (**BB**), Germany (1995)	Sárospatak, Hungary (1995) Visby, Sweden (1995) Soest, Netherlands (2004) Mishawaka, Indiana, USA	Strzelce Opolskie (*Groß Strehlitz*), Poland Riga, Latvia Saint-Parres-aux-Tertres, France

- Spandau, Berlin, amalgamated 1920 into Berlin as the borough of Spandau
- Steinfurt, North Rhine-Westphalia:
  London Borough of Ealing, United Kingdom

- Stendal, Saxony-Anhalt:

  Grenoble, France (1992)

- Stralsund, Mecklenburg-Vorpommern:

Boulogne-sur-Mer, France (1963, not 2012) Pori, Finland (1964) Ventspils (*Windau*), Latvia (1969)	Stargard Szczeciński (*Stargard in Pommern*), Poland (1987) Kiel (**SH**), Germany (1987) Malmö, Sweden (1991)	Svendborg, Denmark (1992) Trelleborg, Sweden (2000)

- Straubing, Bavaria:

Romans-sur-Isère, France (1971)	Wels, Austria (1972)	Tuam, Ireland (1992)

- Stadtbergen, Bavaria:

Fukushima Prefecture, Japan (1974) Baguida, Togo (1986)	Brie-Comte-Robert, France (1987) Olbernhau (**SX**), Germany (1989)	Bagnolo Mella, Italy (2004) Litvínov, Czech Republic

- Stuttgart, Baden-Württemberg:

St Helens, United Kingdom (1948) Cardiff, United Kingdom (1955) St. Louis, Missouri, USA (1960) Strasbourg, France (1962) Mumbai, India (1968)	Menzel Bourguiba, Tunisia (1971) Cairo, Egypt (1979) Łódź, Poland (1988) Ōgaki, Japan (1988) Brno, Czech Republic (1989)	Samara, Russia (1992) Mexico City, Mexico (2006) Shavei Zion, Israel Nanjing, China

twinned with Zuffenhausen:	La Ferté-sous-Jouarre, France (1977)

twinned with Vaihingen:	Melun, France (1985)

twinned with Bad Cannstatt:	Újbuda (district of Budapest), Hungary (1996)

T

- Tauberbischofsheim, Baden-Württemberg:

Duderstadt (**LS**), Germany (1961)	Vitry-le-François, France (1966)

- Tecklenburg, North Rhine-Westphalia:

  Chalonnes-sur-Loire, France (1982)

- Telgte, North Rhine-Westphalia:

Polanica Zdroj, Poland (1995)	Stupino, Russia (1995)	Tomball, Texas, USA (2000)

- Trier, Rhineland-Palatinate:

Gloucester, United Kingdom (1957) Metz, France (1957)	's-Hertogenbosch, Netherlands (1968) Pula, Croatia (1970)	Fort Worth, Texas, USA (1987) Xiamen, China (2010)

- Trossingen, Baden-Württemberg:

Cluses, France (1972)	Beaverton, Oregon, USA (1997)	Windhoek, Namibia (2000)

- Tübingen, Baden-Württemberg:

Monthey, Switzerland (1959) Aix-en-Provence, France (1960) Ann Arbor, Michigan, USA (1965)	Durham, United Kingdom (1969) Aigle, Switzerland (1973) Petrozavodsk, Russia (1989)	Perugia, Italy (1994) Villa El Salvador, Peru (2006)

- Tuttlingen, Baden-Württemberg:

Bex, Switzerland (1979)	Draguignan, France (1989)

U

- Ulmen, Rhineland-Palatinate:

  Lormes, France (1996)

- Unna, North Rhine-Westphalia:

Waalwijk, Netherlands (1968) Talaiseau, France (1969)	Ajka, Hungary (2000)	Kirklees, United Kingdom (2000)

V

- Vechta, Lower Saxony:

Le Cellier, France (1989) Seguin, Texas, USA (1995) Jászberény, Hungary (1993)	Ile-de-Batz, France (2007) Plouénan, France (2007) Plougoulm, France (2007)	Roscoff, France (2007) Santec, France (2007) Sibiril, France (2007)

- Verden, Lower Saxony:

Bagrationovsk, Russia (1966) Górowo Iławeckie, Poland (1996)	Saumur, France (1967) Warwick, United Kingdom (1990)	Zielona Góra, Poland (1993)

- Viersen, North Rhine-Westphalia:

Lambersart, France (1970) Peterborough, United Kingdom (1982)	Kaniv, Ukraine (1996)	Pardesiya, Israel (2001)

W

- Walldorf, Baden-Württemberg:

Astoria, Oregon, USA (1963) Kirklareli, Turkey (1970)	Saint-Max, France (1985) Waldorf, Maryland, USA (2001)	Freeport, New York, USA (2003) Walldorf (**TH**), Germany (2007)

- Warendorf, North Rhine-Westphalia:

Barentin, France (1965) Pavilly, France (1972 with Freckenhorst)	Oleśnica (*Oels*), Poland (2002) Petersfield, England (2007)	Navaro, Italy

- Warstein, North Rhine-Westphalia:

Saint-Pol-sur-Ternoise, France (1964) Wurzen (**SX**), Germany (1990)	Hebden Royd, United Kingdom (1995)	Pietrapaola, Italy (2001)

- Wedel, Schleswig-Holstein:

District of Makete, Tanzania (1982) Caudry, France (1985)	Vejen, Denmark (until 2010)	Wolgast (**MV**), Germany

- Weimar, Thuringia:

Hämeenlinna, Finland (1970) Trier (**RP**), Germany (1987) Siena, Italy (1994)	Blois, France (1995) Shiraz, Iran (2009) Zamość (*Zamosch*), Poland (2012)	Weimar, Texas, USA Fulda (**HE**), Germany

- Wegberg, North Rhine-Westphalia:

Echt-Susteren, Netherlands	Maaseik, Belgium

- Westerburg, Rhineland-Palatinate:

Daventry, United Kingdom Le Cateau-Cambrésis, France	Nowa Wieś Wielka (*Groß Neudorf*), Poland Złotoryja (*Goldberg*), Poland	Grünstadt (**RP**), Germany

- Wetzlar, Hesse:

borough of Neukölln, Berlin (**BE**), Germany (1959) Avignon, France (1960) Colchester, England (1969)	Schladming, Austria (1974) Písek, Czech Republic (1981) Siena, Italy (1987)	Ilmenau (**TH**), Germany (1990)

twinned with district Garbenheim:	Reith, Austria (1976)

- Wiesbaden, Hesse:

Klagenfurt, Austria (1930) Montreux, Switzerland (1953) borough of Friedrichshain-Kreuzberg, Berlin (**BE**), Germany (1964/2001) Ghent, Belgium (1969)	Ljubljana, Slovenia (1977) Kfar Saba, Israel (1981) San Sebastián, Spain (1981) Wrocław, Poland (1987)	Royal Tunbridge Wells, United Kingdom (1989) Görlitz (**SX**), Germany (1990) Ocotal, Nicaragua (1990)

twinned with district Naurod:	Fondettes, France (1975)

- Wilhelmsfeld, Baden-Württemberg:

  Calamba, Philippines (2008)

- Willich, North Rhine-Westphalia:

Linselles, France (1970) Zogoré, Burkina Faso (1985) Bürgel (**TH**), Germany	Newmarket, England (this one?) Premnitz (**BB**), Germany Žalec (*Sachsenfeld*), Slovenia	Smiltene (*Smilten*), Latvia

- Winterlingen, Baden-Württemberg:

Izbica, Poland (2008)	Shiner, Texas, USA

- Wismar, Mecklenburg-Vorpommern:

Kemi, Finland (1959) Aalborg, Denmark (1961) Calais, France (1966)	Lübeck (**SH**), Germany (1987) Halden, Norway (1991) Kalmar, Sweden (2002)	Moss, Norway

- Wolfsburg, Lower Saxony:

Luton, England (1950) Marignane, France (1963) Province of Pesaro e Urbino, Italy (1975)	Sarajevo, Bosnia and Herzegovina (1985) Halberstadt (**SA**), Germany (1989) Togliatti, Russia (1991)	Bielsko-Biała, Poland (1998

- Wolgast, Mecklenburg-Vorpommern:

Nexø, Denmark Sölvesborg, Sweden (1991)	Wedel (**SH**), Germany	Karlino (*Körlin an der Persante*), Poland (2006)

- Worms, Rhineland-Palatinate:

St. Albans, United Kingdom (1957) Auxerre, France (1968) Parma, Italy (1984)	Tiberias, Israel (1986) Bautzen (**SX**), Germany (1990) Mobile, Alabama, USA (1998)	Kivumu, Rwanda

twinned with Pfeddersheim:	Nolay, France (1966)

twinned with Heppenheim:	Ampilly-le-Sec, France (1968)

twinned with Ibersheim:	Chemellier, France (2006)

- Wuppertal, North Rhine-Westphalia:

South Tyneside, United Kingdom (1951) Saint-Étienne, France (1960) borough of Tempelhof-Schöneberg, Berlin (**BE**), Germany (1964)	Beersheba, Israel (1977) Košice (*Kaschau*), Slovakia (1980) Schwerin (**MV**), Germany (1987)	Matagalpa, Nicaragua (1987) Legnica (*Liegnitz*), Poland (1993) Yekaterinburg, Russia (1993)

- Würzburg, Bavaria:

Caen, France (1962) Dundee, United Kingdom (1962) Rochester, New York, USA (1964) Mwanza, Tanzania (1966)	Ōtsu, Japan (1979) Salamanca, Spain (1980) Suhl (**TH**), Germany (1988) Umeå, Sweden (1992)	Trutnov, Czech Republic (1998) Bray and County Wicklow, Ireland (1998)

X

- Xanten, North Rhine-Westphalia:

Geel, Belgium (1990)	Saintes, France (2002)	Salisbury, United Kingdom (2006)

Z

- Zittau, Saxony:

Villingen-Schwenningen (**BW**), Germany (1990) Liberec (*Reichenberg*), Czech Republic	Bogatynia (*Reichenau*), Poland Pistoia, Italy	Portsmouth, Ohio, USA

twinned with district Eichgraben:	Eichgraben, Austria

- Zülpich, North Rhine-Westphalia:

Blaye, France (1972)	Kangasala, Finland (1978)

- Zweibrücken, Rhineland-Palatinate:

Boulogne-sur-Mer, France (1959)	Yorktown, Virginia, USA (1978)	Barrie, Ontario, Canada (1996)

- Zwickau, Saxony:

Jablonec nad Nisou (*Gablonz an der Neiße*), Czech Republic (1971)	Zaanstad, Netherlands (1987)	Dortmund (**NRW**), Germany (1988)

References

[1] "Douzelage.org: Member Towns" (http://www.douzelage.org/index.php?id=15). www.douzelage.org. . Retrieved 21 October 2009.

[2] "Douzelage.org: Home" (http://www.douzelage.org/). www.douzelage.org. . Retrieved 21 October 2009.

[3] "Cities Twinned with Duisburg" (http://www.duisburg.de/micro/english/introducing/102010100000187829.php). www.duisburg.de. . Retrieved 7 May 2009.

[4] "List of Twin Towns in the Ruhr District" (http://www.webcitation.org/5lctsW5KG). © 2009 Twins2010.com (http://www.twins2010.com/index.php?id=home&L=1). Archived from the original (http://www.twins2010.com/fileadmin/user_upload/pic/Dokumente/List_of_Twin_Towns_01.pdf?PHPSESSID=2edd34819db21e450d3bb625549ce4fd) on 28 November 2009. . Retrieved 28 October 2009.

[5] Portsmouth City Council. *Twinning* (http://www.portsmouth.gov.uk/yourcouncil/1396.html). Retrieved 22 August 2007.

[6] Portsmouth Duisburg Anglo-German Friends (http://portsmouth-duisburg.tripod.com/)

[7] Duisburger Portsmouthfreunde (http://www.portsmouthfreunde.de/)

[8] http://twinning.rjnet.co.uk/

[9] "Bristol City – Town twinning" (http://www.bristol.gov.uk/ccm/navigation/leisure-and-culture/tourism-and-travel/town-twinning/). © 2009 Bristol City Council (http://www.bristol.gov.uk/ccm/portal/). . Retrieved 17 July 2009.

[10] *"Poznań Official Website – Twin Towns"* (http://www.poznan.pl/mim/public/publikacje/pages.html?co=list&id=19&ch=20&instance=1017&lang=pl). *© 1998–2008 Urząd Miasta Poznania. . Retrieved 29 November 2008.*

[11] "" (http://www.city.hiroshima.jp/shimin/kokusai/shimai/top-e.html). City.hiroshima.jp. . Retrieved 17 July 2009.

[12] Hannover.de (http://www.hannover.de/data/download/lhh/kultur/japan-wochen-screen.pdf)

[13] "Leipzig – International Relations" (http://www.leipzig.de/int/en/int_messen/partnerstaedte/). © 2009 Leipzig City Council, Office for European and International Affairs. . Retrieved 17 July 2009.

[14] "Partnerstädte Hechingens" (http://www.hechingen.de/default.php?content=764). . Retrieved 23 April 2010.

[15] Binjakëzime (http://www.kamza.gov.al/previewdoc.php?file_id=336), Municipality of Kamëz (in Albanian)

[16] City of Davenport, Iowa | Sister Cities (http://www.cityofdavenportiowa.com/department/board.asp?fDD=5-33)

Bavaria

Bavaria
Freistaat Bayern

— State of Germany —

Flag

Coat of arms

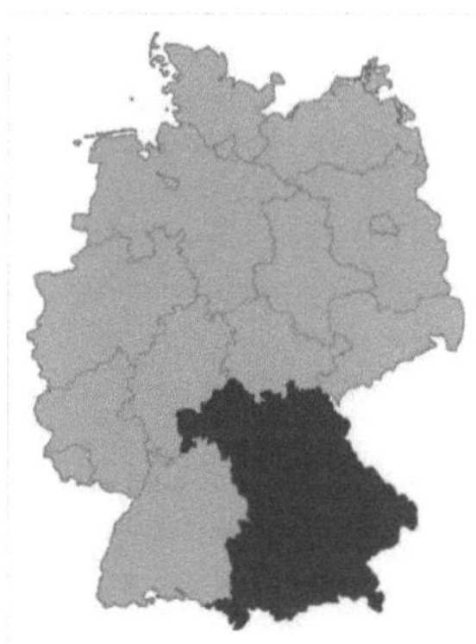

Coordinates: 48°46′39″N 11°25′52″E

Country	Germany
Capital	Munich
Government	
• **Minister-President**	Horst Seehofer (CSU)
• **Governing parties**	CSU / FDP
• **Votes in Bundesrat**	6 (of 69)
Area	
• **Total**	70549.44 km^2 (**unknown operator: u'strong'** sq mi)
Population (2010-11)[1]	
• **Total**	12,542,365
Time zone	CET (UTC+1)
• **Summer (DST)**	CEST (UTC+2)

ISO 3166 code	DE-BY
GDP/ Nominal	€446.44 billion (2011) [2]
GDP per capita	€35,595 (2011)
NUTS Region	DE2
Website	bayern.de [3]

The **Free State of Bavaria** (German: *Freistaat Bayern*, pronounced [ˈfʁaɪʃtaːt ˈbaɪ.ɐn] (listen)), is a state of Germany, located in the southeast of Germany. With an area of 70548 square kilometres (27200 sq mi), it is the largest state by area, forming almost 20% of the total land area of Germany. Bavaria is Germany's second most populous state (after North Rhine-Westphalia), with 12.5 million inhabitants, more than any of the three sovereign nations on its borders. Bavaria's capital and largest city is Munich, the third largest city in Germany.

One of the oldest states of Europe, it was established as a duchy in the mid first millennium. In the 17th century, the Duke of Bavaria became a Prince-elector of the Holy Roman Empire of the German Nation. The Kingdom of Bavaria existed from 1806 to 1918, and Bavaria has since been a free state (republic). Modern Bavaria also includes parts of the historical regions of Franconia, Upper Palatinate and Swabia.

History

Prehistoric *Heunischenburg*, in the vicinity of Kronach

The Bavarians emerged in a region north of the Alps, originally inhabited by the Gauls, which had been part of the Roman provinces of Raetia and Noricum. The Bavarians spoke Old High German but, unlike other Germanic groups, probably did not migrate from elsewhere. Rather, they seem to have coalesced out of other groups left behind by Roman withdrawal late in the 5th century. These peoples may have included the Celtic Boii, some remaining Romans, Marcomanni, Allemanni, Quadi, Thuringians, Goths, Scirians, Rugians, Heruli. The name "Bavarian" ("Baiuvarii") means "Men of Baia" which may indicate Bohemia, the homeland of the Celtic Boii and later of the Marcomanni. They first appear in written sources c. 520. Saint Boniface completed the people's conversion to Christianity in the early-8th century. Bavaria was, for the most part, unaffected by the Protestant Reformation that happened centuries later.

From about 554 to 788, the house of Agilolfing ruled the Duchy of Bavaria, ending with Tassilo III who was deposed by Charlemagne.

Three early dukes are named in Frankish sources: Garibald I may have been appointed to the office by the Merovingian kings and married the Lombard princess Walderada when the church forbade her to King Chlothar I in 555. Their daughter, Theodelinde, became Queen of the Lombards in northern Italy and Garibald was forced to flee to her when he fell out with his Frankish overlords. Garibald's successor, Tassilo I, tried unsuccessfully to hold the eastern frontier against the expansion of Slavs and Avars around 600. Tassilo's son Garibald II seems to have achieved a balance of power between 610 and 616.

After Garibald II little is known of the Bavarians until Duke Theodo I, whose reign may have begun as early as 680. From 696 onwards he invited churchmen from the west to organize churches and strengthen Christianity in his duchy (it is unclear what Bavarian religious life consisted of before this time). His son, Theudebert, led a decisive Bavarian campaign to intervene in a succession dispute in the Lombard Kingdom in 714, and married his sister Guntrud to the Lombard King Liutprand. At Theodo's death the duchy was divided among his sons, but reunited under his grandson

Hucbert.

At Hucbert's death (735) the duchy passed to a distant relative named Odilo, from neighbouring Alemannia (modern southwest Germany and northern Switzerland). Odilo issued a law code for Bavaria, completed the process of church organisation in partnership with St. Boniface (739), and tried to intervene in Frankish succession disputes by fighting for the claims of the Carolingian Grifo. He was defeated near Augsburg in 743 but continued to rule until his death in 748.

Middle Ages

Further information: Duchy of Bavaria

Tassilo III (b. 741 - d. after 794) succeeded his father at the age of eight after an unsuccessful attempt by Grifo to rule Bavaria. He initially ruled under Frankish oversight but began to function independently from 763 onwards. He was particularly noted for founding new monasteries and for expanding eastwards, fighting Slavs in the eastern Alps and along the River Danube and colonising these lands. After 781, however, his cousin Charlemagne began to pressure Tassilo to submit and finally deposed him in 788. The deposition was not entirely legitimate. Dissenters attempted a coup against Charlemagne at Tassilo's old capital of Regensburg in 792, led by his own son Pépin the Hunchback, and the king had to drag Tassilo out of imprisonment to formally renounce his rights and titles at the Assembly of Frankfurt in 794. This is the last appearance of Tassilo in the sources and he probably died a monk. As all of his family were also forced into monasteries, this was the end of the Agilolfing dynasty.

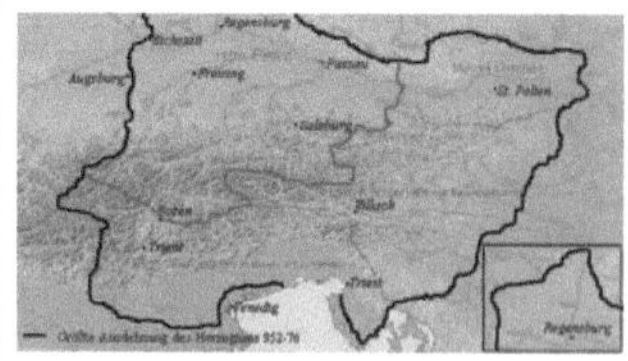

Bavaria in the 10th century

For the next 400 years numerous families held the duchy, rarely for more than three generations. With the revolt of duke Henry the Quarrelsome in 976, Bavaria lost large territories in the south and south east. Among them a mark called "Ostarrichi" which was elevated to a duchy out of own right and given to the Babenberger family. This event marks the birth of Austria. The last, and one of the most important, of these dukes was Henry the Lion of the house of Welf, founder of Munich, de facto the second most powerful man in the empire as the ruler of two duchies. When in 1180, Henry the Lion was deposed as Duke of Saxony and Bavaria by his cousin, Frederick I, Holy Roman Emperor (aka "Barbarossa" for his red beard), Bavaria was awarded as fief to the Wittelsbach family, counts palatinate of Schyren ("Scheyern" in modern German), which ruled from 1180 to 1918. The Electorate of the Palatinate by Rhine ("Kurpfalz" in German) was also acquired by the House of Wittelsbach in 1214.

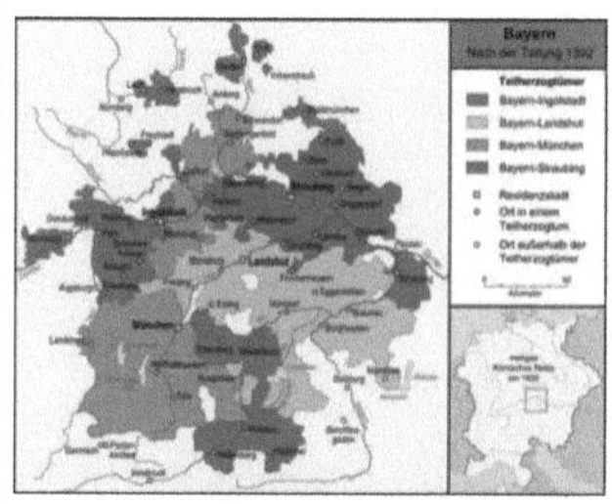

Bavarian duchies after the partition of 1392

The first of several divisions of the duchy of Bavaria occurred in 1255. With the extinction of the Hohenstaufen in 1268 also Swabian territories were acquired by the Wittelsbach dukes. Emperor Louis the Bavarian acquired Brandenburg, Tirol, Holland and Hainaut for his House but released the Upper Palatinate for the Palatinate branch of the Wittelsbach in 1329. In 1506 with the Landshut War of Succession the other parts of Bavaria were reunited and Munich became the sole capital.

17th and 18th centuries

Further information: Electorate of Bavaria

In 1623 the Bavarian duke replaced his relative of the Palatinate branch, the Electorate of the Palatinate in the early days of the Thirty Years' War and acquired the powerful prince-electoral dignity in the Holy Roman Empire, determining its Emperor thence forward, as well as special legal status under the empire's laws. The country became one of the Jesuit supported counter-reformation centers. During the early and mid-18th century the ambitions of the Bavarian prince electors led to several wars with Austria as well as occupations by Austria (Spanish succession, election of a Wittelsbach emperor instead of a Habsburger). From 1777 onwards and after the old Bavarian branch of the family had died out with elector Max III Joseph, Bavaria and the Electorate of the Palatinate were governed once again in personal union, now by the Palatinian lines.

Kingdom of Bavaria

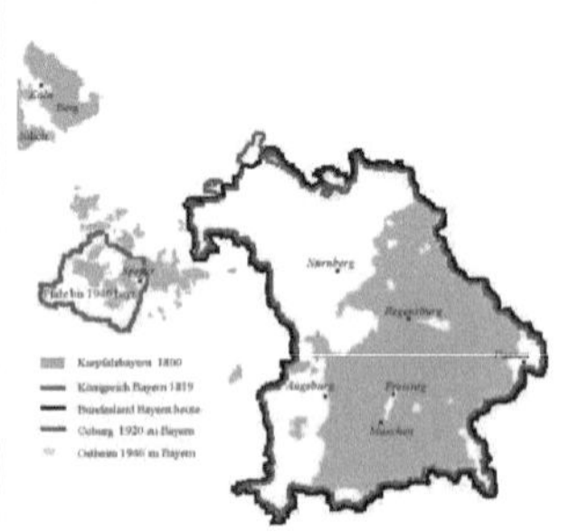

Bavaria in the 19th century and beyond

When Napoleon abolished the Holy Roman Empire, Bavaria became a kingdom in 1806, and its area doubled. Tirol was temporarily united, Salzburg temporarily reunited with Bavaria but finally ceded to Austria. In return the Rhenish Palatinate and Franconia were annexed to Bavaria in 1815. Between 1799 and 1817 the leading minister count Montgelas followed a strict policy of modernisation and laid the foundations of administrative structures that survived even the monarchy and are (in their core) valid until today. In 1808 a first and in 1818 a more modern constitution (by the standards of the time) was passed, that established a bicameral Parliament with a House of Lords (*Kammer der Reichsräte*) and a House of Commons (*Kammer der Abgeordneten*). The constitution would last until the collapse of the monarchy at the end of World War I.

Bavaria as a part of the German Empire

Bavarian stamps during the German empire period

After the rise of Prussia to prominence Bavaria managed to preserve its independence by playing off the rivalries of Prussia and Austria. Allied to Austria, it was defeated in the 1866 Austro-Prussian War and did not belong to the North German Federation of 1867, but the question of German unity was still alive. When France attacked Prussia in 1870, the south German states Baden, Württemberg, Hessen-Darmstadt and Bavaria joined the Prussian forces and ultimately joined the Federation, which was renamed *Deutsches Reich* (German Empire) in 1871. Bavaria continued as a monarchy, and it even had some special rights within the federation (such as an army, railways and a postal service of its own).

In the early-20th century Wassily Kandinsky, Paul Klee, Henrik Ibsen, and other notable artists were drawn to Bavaria, notably to the Schwabing district of Munich, later devastated by World War II.

20th century

A memorial to soldiers who died in the two World Wars. Village in Bavaria.

On November 12, 1918, Ludwig III signed a document, the Anif declaration, releasing both civil and military officers from their oaths; the newly-formed republican government of Socialist premier Kurt Eisner interpreted this as an abdication. To date, however, no member of the house of Wittelsbach has ever formally declared renunciation of the throne. On the other hand, none has ever since officially called upon their Bavarian or Stuart claims. Family members are active in cultural and social life, including the head of the house, HRH Duke Franz in Bavaria. They step back from any announcements on public affairs, showing approval or disapproval solely by HRH's presence or absence.

Eisner was assassinated in February 1919 ultimately leading to a Communist revolt and the short-lived Bavarian Socialist Republic being proclaimed 6 April 1919. After violent suppression by elements of the German Army and notably the Freikorps, the Bavarian Socialist Republic fell in May 1919. The Bamberg Constitution (*Bamberger Verfassung*) was enacted on 12 or 14 August 1919 and came into force on 15 September 1919 creating the Free State of Bavaria within the Weimar Republic. Extremist activity further increased, notably the 1923 Beer Hall Putsch led by the National Socialists, and Munich and Nuremberg became Nazi strongholds under the Third Reich. As a manufacturing center, Munich was heavily bombed during World War II and occupied by U.S. troops. The Rhenish Palatinate was detached from Bavaria in 1946 and made part of the new state Rhineland-Palatinate. During the Cold War, Bavaria was part of West Germany.

Since World War II, Bavaria has been rehabilitated from a poor agrarian province into a prosperous industrial hub. A massive reconstruction effort restored much of Munich's and other places historic cores. The state capital hosted the 1972 Summer Olympics and matches of the Football World Cups of 1974 and 2006 as well as European Track & Field championships. More recently, former state minister-president Edmund Stoiber was the CDU/CSU candidate for chancellor in the 2002 federal election which he lost, and native son Cardinal Joseph Ratzinger was elected Pope Benedict XVI in 2005.

Coat of arms

Bavarian herald Joerg Rugenn wearing a tabard of the arms around 1510

The modern coat of arms of Bavaria was designed by Eduard Ege in 1946, following heraldic traditions.

- The Golden Lion: At the dexter chief, sable, a lion rampant Or, armed and langued gules. This represents the administrative region of Upper Palatinate.
- The "Franconian Rake": At the sinister chief, per fess dancetty, gules and argent. This represents the administrative regions of Upper, Middle and Lower Franconia.
- The Blue Panther: At the dexter base, argent, a panther rampant azure, armed Or and langued gules. This represents the regions of Lower and Upper Bavaria.
- The Three Lions: At the sinister base, Or, three lions passant guardant sable, armed and langued gules. This represents Swabia.
- The White-And-Blue inescutcheon: The inescutcheon of white and blue fusils askance was originally the coat of arms of the Counts of Bogen, adopted in 1247 by the Wittelsbachs House. The white-and-blue fusils are indisputably the emblem of Bavaria and these arms today symbolize Bavaria as a whole. Along with the People's Crown, it is officially used as the Minor Coat of Arms.
- The People's Crown: The coat of arms is surmounted by a crown with a golden band inset with precious stones and decorated with five ornamental leaves. This crown first appeared in the coat of arms to symbolize sovereignty of the people after the royal crown was eschewed in 1923.

Geography

The Bavarian Alps

Bavaria shares international borders with Austria and the Czech Republic as well as with Switzerland (across Lake Constance). Because all of these countries are part of the Schengen Area, the border is completely open. Neighbouring states within Germany are Baden-Württemberg, Hesse, Thuringia and Saxony. Two major rivers flow through the state, the Danube (*Donau*) and the Main. The Bavarian Alps define the border with Austria, (including the Austrian federal-states of Vorarlberg, Tyrol and Salzburg) and within the range is the highest peak in Germany, the Zugspitze. The Bavarian Forest and the Bohemian Forest form the vast majority of the frontier with the Czech Republic and Bohemia.

The major cities in Bavaria are Munich (*München*), Nuremberg (*Nürnberg*), Augsburg, Regensburg, Würzburg, Ingolstadt, Fürth and Erlangen.

Population and area

Administrative region	Capital	Population (2011)		Area (km^2)		No. municipalities	
Lower Bavaria	Landshut	1,192,641	9.48%	10,330	14.6%	258	12.5%
Lower Franconia	Würzburg	1,315,882	10.46%	8,531	12.1%	308	15.0%
Upper Franconia	Bayreuth	1,067,988	8.49%	7,231	10.2%	214	10.4%
Middle Franconia	Ansbach	1,717,670	13.65%	7,245	10.3%	210	10.2%
Upper Palatinate	Regensburg	1,081,800	8.60%	9,691	13.7%	226	11.0%
Swabia	Augsburg	1,788,729	14.21%	9,992	14.2%	340	16.5%
Upper Bavaria	Munich	4,418,828	35.12%	17,530	24.8%	500	24.3%
Total		**12,583,538**	**100.0%**	**70,549**	**100.0%**	**2,056**	**100.0%**

Major cities

City	Region	Inhabitants 31 December 2000	Inhabitants 31 December 2005	Inhabitants 31 December 2008
Munich	Upper Bavaria	1,210,223	1,259,677	1,326,807
Nuremberg	Middle Franconia	488,400	499,237	503,638
Augsburg	Swabia	254,982	262,676	263,313
Regensburg	Upper Palatinate	125,676	129,859	133,525
Würzburg	Lower Franconia	127,966	133,906	133,501
Ingolstadt	Upper Bavaria	115,722	121,314	123,925
Fürth	Middle Franconia	110,477	113,422	114,071
Erlangen	Middle Franconia	100,778	103,197	104,980
Bayreuth	Upper Franconia	74,153	73,997	72,935
Bamberg	Upper Franconia	69,036	70,081	69,989
Aschaffenburg	Lower Franconia	67,592	68,642	68,747
Landshut	Lower Bavaria	58,746	61,368	62,606
Kempten (Allgäu)	Swabia	61,389	61,360	62,135
Rosenheim	Upper Bavaria	58,908	60,226	60,711
Neu-Ulm	Swabia	50,188	51,410	53,866
Schweinfurt	Lower Franconia	54,325	54,273	53,588
Passau	Lower Bavaria	50,536	50,651	50,717
Hof	Upper Franconia	50,741	48,723	47,275
Freising	Upper Bavaria	44,167	45,827	45,654
Straubing	Lower Bavaria	44,014	44,633	44,496

Administrative divisions

Regierungsbezirke (administrative districts)

Bavaria is divided into 7 administrative districts called *Regierungsbezirke* (singular *Regierungsbezirk*).

1. Upper Franconia (German: *Oberfranken*)
2. Middle Franconia (*Mittelfranken*)
3. Lower Franconia (*Unterfranken*)
4. Swabia (*Schwaben*)
5. Upper Palatinate (*Oberpfalz*)
6. Upper Bavaria (*Oberbayern*)
7. Lower Bavaria (*Niederbayern*)

Administrative Districts (*Regierungsbezirke* and *Bezirke*) of Bavaria

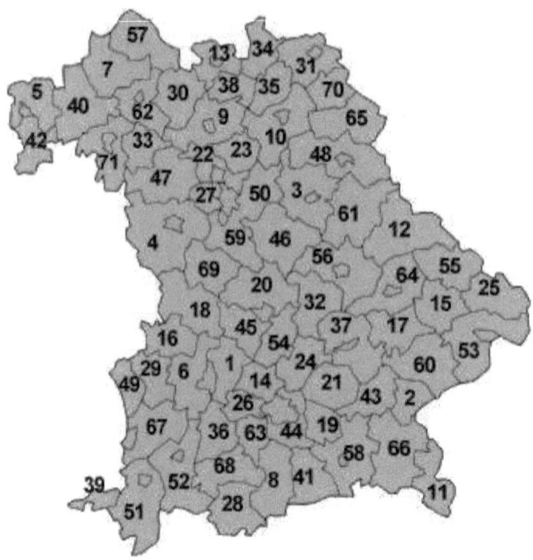

Bezirke

Bezirke (districts) are the third communal layer in Bavaria; the others are the *Landkreise* and the *Gemeinden* or *Städte*. The *Bezirke* in Bavaria are territorially identical with the *Regierungsbezirke* (e.g. Regierung von Oberbayern), but are a different form of administration, having their own parliaments, etc.

In the larger states of Germany (including Bavaria), there are *Regierungsbezirke* which are only administrative divisions and not self-governing entities as the *Bezirke* in Bavaria.

Landkreise/kreisfreie cities

These administrative regions consist of 71 administrative districts (called *Landkreise*, singular *Landkreis*) and 25 independent cities (*kreisfreie Städte*, singular *kreisfreie Stadt*).

Landkreise:

Munich

Nuremberg

Augsburg

Regensburg

Würzburg

Bayreuth

- Aichach-Friedberg
- Altötting
- Amberg-Sulzbach
- Ansbach
- Aschaffenburg
- Augsburg
- Bad Kissingen
- Bad Tölz-Wolfratshausen
- Bamberg
- Bayreuth
- Berchtesgadener Land
- Cham
- Coburg
- Dachau
- Deggendorf
- Dillingen
- Dingolfing-Landau
- Donau-Ries
- Ebersberg
- Eichstätt
- Erding
- Erlangen-Höchstadt
- Forchheim
- Freising
- Freyung-Grafenau
- Fürstenfeldbruck
- Landsberg
- Landshut
- Lichtenfels
- Lindau
- Main-Spessart
- Miesbach
- Miltenberg
- Mühldorf
- München (*Landkreis München*)
- Neuburg-Schrobenhausen
- Neumarkt
- Neustadt (Aisch)-Bad Windsheim
- Neustadt (Waldnaab)
- Neu-Ulm
- Nürnberger Land
- Oberallgäu
- Ostallgäu
- Passau
- Pfaffenhofen
- Regen
- Regensburg
- Rhön-Grabfeld
- Rosenheim
- Roth
- Rottal-Inn
- Schwandorf

- Fürth
- Garmisch-Partenkirchen
- Günzburg
- Hassberge
- Hof
- Kelheim
- Kitzingen
- Kronach
- Kulmbach
- Schweinfurt
- Starnberg
- Straubing-Bogen
- Tirschenreuth
- Traunstein
- Unterallgäu
- Weilheim-Schongau
- Weissenburg-Gunzenhausen
- Wunsiedel
- Würzburg

Kreis-free Cities:

- Amberg
- Ansbach
- Aschaffenburg
- Augsburg
- Bamberg
- Bayreuth
- Coburg
- Erlangen
- Fürth
- Hof
- Ingolstadt
- Kaufbeuren
- Kempten
- Landshut
- Memmingen
- Munich (*München*)
- Nuremberg (*Nürnberg*)
- Passau
- Regensburg
- Rosenheim
- Schwabach
- Schweinfurt
- Straubing
- Weiden
- Würzburg

Gemeinden (municipalities)

The 71 administrative districts are on the lowest level divided into 2031 municipalities (called *Gemeinden*, singular *Gemeinde*). Together with the 25 independent cities (*kreisfreie Städte*, which are in effect municipalities independent of *Landkreis* administrations), there are a total of 2056 municipalities in Bavaria.

In 44 of the 71 administrative districts, there are a total of 215 unincorporated areas (as of January 1, 2005, called *gemeindefreie Gebiete*, singular *gemeindefreies Gebiet*), not belonging to any municipality, all uninhabited, mostly forested areas, but also four lakes (Chiemsee-without islands, Starnberger See-without island Roseninsel, Ammersee, which are the three largest lakes of Bavaria, and Waginger See).

Government and politics

Government

Bavaria has a unicameral *Landtag*, or state parliament, elected by universal suffrage. Until December 1999, there was also a *Senat*, or Senate, whose members were chosen by social and economic groups in Bavaria, but following a referendum in 1998, this institution was abolished.

The Bavarian State Government consists of the Minister-President of Bavaria, 11 Ministers and 6 Secretaries of State. The Minister-President is elected for a period of 5 years by the State Parliament and is head of state. With the approval of the State Parliament he appoints the members of the State Government. The State Government is composed of the:

- Ministry of the Interior (*Staatsministerium des Innern*)
- Ministry of Education and Culture (*Staatsministerium für Unterricht und Kultus*)

- Ministry of Finance (*Staatsministerium der Finanzen*)
- Ministry of Economic Affairs, Infrastructure, Transportation and Technology (*Staatsministerium für Wirtschaft, Infrastruktur, Verkehr und Technologie*, MWIVT)
- Ministry of Environment and Health (*Staatsministerium für Umwelt und Gesundheit*)
- Ministry of Labour and Social Affairs, Family and Women (*Staatsministerium für Arbeit und Sozialordnung, Familie und Frauen*)
- Ministry of Justice and Consumer Protection (*Staatsministerium der Justiz und für Verbraucherschutz*)
- Ministry of Science, Research and Art (*Staatsministerium für Wissenschaft, Forschung und Kunst*, MWFK)
- Ministry of Food, Agriculture and Forestry (*Staatsministerium für Ernährung, Landwirtschaft und Forsten*)
- Ministry of Federal and European Affairs (*Staatsminister für Bundes- und Europaangelegenheiten*)

Political processes also take place in the 7 regions (*Regierungsbezirke* / *Bezirke*) in Bavaria, in the 71 administrative districts (*Landkreise*) and the 25 towns and cities forming their own districts (*kreisfreie Städte*), and in the 2,031 local authorities (*Gemeinden*).

In 1995 Bavaria introduced direct democracy on the local level in a referendum. This was initiated bottom-up by an association called *Mehr Demokratie* (More Democracy). This is a grass-roots organization which campaigns for the right to citizen-initiated referendums. In 1997 the Bavarian Supreme Court aggravated the regulations considerably (e.g. by introducing a turn-out quorum). Nevertheless, Bavaria has the most advanced regulations on local direct democracy in Germany. This has led to a spirited citizens' participation in communal and municipal affairs—835 referenda took place from 1995 through 2005.

Politics

Bavaria has a multi-party system where the biggest parties are the conservative Christian Social Union of Bavaria (CSU), which has dominated politics since 1945 and won every election since then, and the center-left Social Democratic Party of Germany (SPD). The German green party, Alliance '90/The Greens is represented in the parliament as well. Since 2008 Germany's liberal party, the Free Democratic Party (FDP) and the Free Voters are represented in Bavaria's parliament as well. CSU and FDP agreed in October 2008 to form a coalition, while SPD, Free Voters and the Greens form the opposition.

12th and current Minister-President of Bavaria Horst Seehofer

In the 2003 elections the CSU won more than two thirds of the seats in Landtag—something no party had ever achieved in post-war German history. In the following 2008 elections the CSU lost its absolute majority in the Landtag for the first time in 46 years.[4]

This loss is probably attributed to its push for an anti-smoking law, the most stringent in Germany, which became one of the most controversial laws ever enacted in Bavaria. As result, the CSU changed its stance and weakened the anti-smoking law to allow some more loopholes. However, the citizens held a successful petition drive in November–December 2009 to call for a total smoking ban. The CSU dismissed the petition and demands as unnecessary and frivolous, but the referendum proceeded and the voters voted for the smoking ban.

Minister-presidents of Bavaria since 1945

Minister-presidents of Bavaria					
No.	Name	Born-Died	Party affiliation	Begin of Tenure	End of Tenure
1	Fritz Schäffer	1888–1967	CSU	1945	1945
2	Wilhelm Hoegner	1887–1980	SPD	1945	1946
3	Hans Ehard	1887–1980	CSU	1946	1954
4	Wilhelm Hoegner	1887–1980	SPD	1954	1957
5	Hanns Seidel	1901–1961	CSU	1957	1960
6	Hans Ehard	1887–1980	CSU	1960	1962
7	Alfons Goppel	1905–1991	CSU	1962	1978
8	Franz Josef Strauß	1915–1988	CSU	1978	1988
9	Max Streibl	1932–1998	CSU	1988	1993
10	Edmund Stoiber	*1941	CSU	1993	2007
11	Günther Beckstein	*1943	CSU	2007	2008
12	Horst Seehofer	*1949	CSU	2008	incumbent

Economy

Bavaria has long had one of the largest and healthiest economies of any region in Germany, or Europe for that matter.[5] Its GDP in 2007 exceeded 434 billion Euros (about 600 bn US$).[6] This makes Bavaria itself one of the largest economies in Europe and only 17 countries in the world have higher GDP.[7] Some large companies headquartered in Bavaria include BMW, Siemens, Rohde & Schwarz, Audi, Munich Re, Allianz, Infineon, MAN, Wacker Chemie, Puma, and Adidas. Bavaria has a GDP per capita of over $48 000 US, meaning that if it were its own independent country it would rank 7th or 8th in the world.

Company names

The motorcycle and automobile makers BMW (*Bayerische Motoren-Werke*, or Bavarian Motor Works) and Audi, Allianz, Grundig (consumer electronics), Siemens (electricity, telephones, informatics, medical instruments), Continental (Automotive Tire and Electronics), Adidas, Puma, HypoVereinsbank (UniCredit Group), Infineon and Krauss-Maffei Wegmann have (or had) a Bavarian industrial base.

Bavaria has also given its name to the largest Colombian brewery (Cervecería Bavaria) and a major Dutch brewery (Bavaria Bier).

Culture

Religion in Bavaria - 2010

religion	percent
Roman Catholics	54.4%
Protestants	20.4%
Muslims	4.0%
Other or none	21.2%

Some features of the Bavarian culture and mentality are remarkably distinct from the rest of Germany. Noteworthy differences (especially in rural areas, less significant in the major cities) can be found with respect to:

Bavarian church with Alps in the background

Religion

Bavarian culture (*Altbayern*) has a long and predominant tradition of Roman Catholic faith. The current pope, Benedict XVI (Joseph Alois Ratzinger), was born in Marktl am Inn in Upper Bavaria and was Cardinal-Archbishop of Munich and Freising. Otherwise, the culturally Franconian and Swabian regions of the modern State of Bavaria are historically more diverse in religiosity, with both Catholic and Protestant traditions.

As of 2010 54.4% of Bavarians still adhere to Roman Catholicism though the number is on the decline (they were 70.4% in 1970, 56.3% in 2007).[8] 20,4% of the population adheres to the Evangelical Lutheran Church in Bavaria, and their number is declining too.[8] Muslims make up 4.0% of the population of Bavaria.[9] 21.2% of Bavarians are irreligious or adhere to other religions.

Though only a relatively small part belongs to the Alps, the perception of Bavaria as an alpine region endures.

Traditions

Bavarians commonly emphasize pride in their traditions. Traditional costumes collectively known as Tracht are worn on special occasions and include in Altbayern Lederhosen for males and Dirndl for females. Centuries-old folk music is performed. The Maibaum, or Maypole (which in the Middle Ages served as the community's yellow pages, as figurettes on the pole represent the trades of the village), and the bagpipes in the Upper Palatinate region bear witness to the ancient Celtic and Germanic remnants of cultural heritage of the region. There are a lot of traditional Bavarian sports disciplines, e.g. the Aperschnalzen is an old tradition of competitive whipcracking.

Whether actually in Bavaria, overseas or full of citizens from other nations they continue to cultivate their traditions. They hold festivals and dances to keep their traditions alive. In New York the German American Cultural Society is a larger umbrella group for others such as the Bavarian organizations, which represent a specific part of Germany. They proudly put forth a German Parade called Steuben Parade each year. Various affiliated events take place amongst its groups, one of which is the Bavarian Dancers.

Food and drink

Bavarians tend to place a great value on food and drink. In addition to their renowned dishes, Bavarians also consume many items of food and drink which are unusual elsewhere in Germany; for example Weisswurst ("white sausage") or in some instances a variety of entrails. At folk festivals and in many beer gardens, beer is traditionally served by the litre (the so-called Maß). Bavarians are particularly proud of the traditional Reinheitsgebot, or purity law, initially established by the Duke of Bavaria for the City of Munich (e.g. the court) in 1487 and the duchy in 1516. According to this law, only three ingredients were allowed in beer: water, barley, and hops. In 1906 the Reinheitsgebot made its way to all-German law, and remained a law in Germany until the EU struck it down recently as incompatible with the European common market. German breweries, however, cling to the principle. Bavarians are also known as some of the world's most beer-loving people with an average annual consumption of 170 litres per person, although figures have been declining in recent years.

Bavaria is also home to the Franconia wine region, which is situated along the Main River in Franconia. The region has produced wine (*Frankenwein*) for over 1,000 years and is famous for its use of the Bocksbeutel wine bottle. The production of wine forms an integral part of the regional culture, and many of its villages and cities hold their own wine festivals (Weinfeste) throughout the year.

Language and dialects

Three German dialects are spoken in Bavaria: Austro-Bavarian in Old Bavaria (South-East and East), Swabian German (an Alemannic German dialect) in the Bavarian part of Swabia (South West) and East Franconian German in Franconia (North).

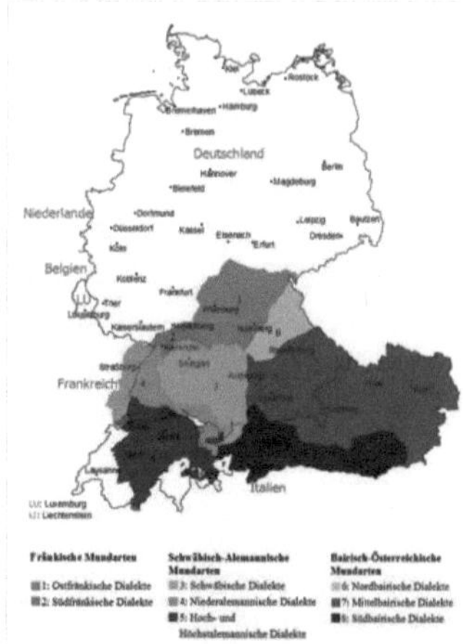

Upper German, southern counterpart to Central German, both forming the High German Languages. Blue are the Austro-Bavarian dialects

Ethnography

Bavarians consider themselves to be egalitarian and informal. Their sociability can be experienced at the annual Oktoberfest, the world's largest beer festival, which welcomes around six million visitors every year, or in the famous beer gardens. In traditional Bavarian beer gardens, patrons may bring their own food but only buy beer from the brewery that runs the beer garden.[10]

In the United States, particularly among German Americans, Bavarian culture is viewed somewhat nostalgically, and many "Bavarian villages", most notably Frankenmuth, Michigan, Helen, Georgia and Leavenworth, Washington, have been founded. Since 1962, the latter has been styled with a Bavarian theme; it is also home to "one of the world's largest collections of nutcrackers" and an Oktoberfest celebration it claims is among the most attended in the world outside of Munich.[11]

Sports

Football

Bavaria is home to several football clubs including FC Bayern Munich, 1. FC Nuremberg, FC Augsburg, TSV 1860 München, and SpVgg Greuther Fürth.

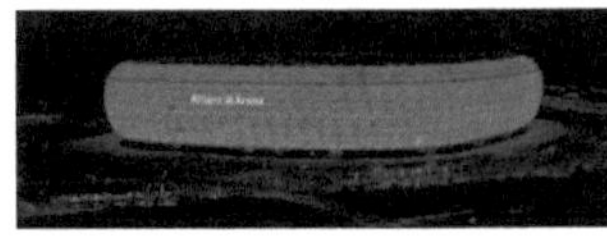

The Allianz Arena

Historical buildings

Johannisburg Castle in Aschaffenburg

Würzburg Residence

Fortress Marienberg and the Alte Mainbrücke in Würzburg

Plassenburg Castle in Kulmbach

Cathedral in Bamberg

Basilica of the Vierzehnheiligen

Castle of Coburg

Festspielhaus of Richard Wagner in Bayreuth

Imperial Castle in Nuremberg

Kastell Biriciana, Weissenburg close to the Limes

Kreuztor in Ingolstadt

Castle of Neuburg an der Donau

Old Stone Bridge and Cathedral of Regensburg

Walhalla temple in Donaustauf near Regensburg

Befreiungshalle in Kelheim

Cathedral and Oberhaus fortification in Passau

Trausnitz castle, Landshut

Burghausen Castle

Townhall in Augsburg

Frauenkirche in Munich

Residenz in Munich

Nymphenburg Palace in Munich

Cathedral in Freising

Herrenchiemsee Palace

Linderhof Palace

Hohenschwangau Castle

Wieskirche, Steingaden

Church of St. Bartholomew at Königssee

Neuschwanstein

Neuschwanstein was built for King Ludwig II, as a second home. It remains unfinished.

Neuschwanstein Castle

Famous people

There are many famous people who were born or lived in present-day Bavaria:

- **Popes** Pope Benedict XVI (baptismal name: **Joseph Aloisius Ratzinger**)—the current Pope of the Roman Catholic Church; Pope Damasus II and Pope Victor II.
- **Painters** such as Hans Holbein the Elder, Albrecht Dürer, Albrecht Altdorfer, Lucas Cranach, Carl Spitzweg, Franz von Lenbach, Franz von Stuck, Franz Marc, Paul Klee, Erwin Eisch, Gabriele Münter.
- **Musicians** such as Johannes Heesters, Orlando di Lasso, Christoph Willibald Gluck, Richard Wagner (originally from Saxony), Richard Strauss, Carl Orff, Johann Pachelbel and Theobald Boehm, the inventor of the modern flute, and countertenor Klaus Nomi.
    - Modern musicians Klaus Doldinger, Barbara Dennerlein, Hans-Jürgen Buchner, Sportfreunde Stiller.
- Opera singers like Jonas Kaufmann and Diana Damrau.
- **Writers**, **poets** and **playwrights** like Hans Sachs, Jean Paul, Frank Wedekind, Christian Morgenstern, Oskar Maria Graf, Bertolt Brecht, Lion Feuchtwanger, Thomas Mann and his sons Klaus and Golo Mann, Ludwig Thoma, Henrik Ibsen, Michael Ende.

- **Scientists** such as Max Planck, Wilhelm Conrad Röntgen, and Werner Heisenberg, as well as Adam Ries, Joseph von Fraunhofer, Georg Ohm, Johannes Stark, Carl von Linde, Ludwig Prandtl, Rudolf Moessbauer, Lothar Rohde, Hermann Schwarz, Robert Huber, Martin Behaim, Levi Strauss, and Rudolf Diesel.
- **Physicians** like Max Joseph von Pettenkofer, Sebastian Kneipp and the neurologist Alois Alzheimer, who first described Alzheimer's Disease.
- **Football players** like Max Morlock, Karl Mai, Franz Beckenbauer, Sepp Maier, Gerd Müller, Paul Breitner, Bernd Schuster, Klaus Augenthaler, Lothar Matthäus, Philipp Lahm, Bastian Schweinsteiger, Holger Badstuber, Thomas Müller, Dietmar Hamann and Stefan Reuter
- **Other sportspeople** such as golfer Bernhard Langer and basketball player Dirk Nowitzki
- **Actors** like Werner Stocker, Helmut Fischer, Walter Sedlmayr, Gustl Bayrhammer, Ottfried Fischer, Ruth Drexel, Elmar Wepper, Fritz Wepper, Uschi Glas, Yank Azman.
- **Entertainers** Siegfried Fischbacher
- **Film directors** Rainer Werner Fassbinder, Bernd Eichinger, Joseph Vilsmaier, and Werner Herzog.
- **Mysterious people**: Kaspar Hauser (the famous foundling), The Smith of Kochel (legend).
- **Legendary outlaws** such as Mathias Kneißl, the legendary robber or Matthias Klostermayr, better known as Bavarian Hiasl.
- Famous army officer, aristocrat and co-conspirator of the 20 July plot Claus von Stauffenberg, born in early 20th century Swabia.
- Noted automobile designer Peter Schreyer, born in Bad Reichenhall
- The fashion designer Damir Doma, who shows men's and women's collection in Paris, grew up in Traunstein.

See also

- Outline of Germany
- List of rulers of Bavaria
- List of Premiers of Bavaria
- Former countries in Europe after 1815
- Extensive pictures of Bavaria in addition to those shown below are linked from in Category:Bavaria, where they are organized (predominantly) by locale.

References

[1] "State population" (http://www.statistik-portal.de/Statistik-Portal/de_zs01_by.asp). *Portal of the Federal Statistics Office Germany*. August 2011. Archived (http://web.archive.org/web/20110524080737/http://www.statistik-portal.de/Statistik-Portal/de_zs01_by.asp) from the original on 24 May 2011. . Retrieved 2011-04-17.

[2] "State GDP" (http://www.statistik-portal.de/statistik-portal/en/en_jb27_jahrtab65.asp). *Portal of the Federal Statistics Office Germany*. . Retrieved 2012-08-27.

[3] http://www.bayern.de/English-.594.htm

[4] n-tv:Fiasko für die CSU (http://www.n-tv.de/Koalitionspartner_noetig_Fiasko_fuer_die_CSU/280920085818/1030029.html)

[5] Its GDP is 143% of the EU average (as of 2005) against a German average of 121.5%, see Eurostat (http://epp.eurostat.ec.europa.eu/extraction/evalight/EVAlight.jsp?A=1&language=en&root=/theme1/reg/reg_e2gdp)

[6] Gemeinsames Datenangebot der Statistischen Ämter des Bundes und der Länder (http://www.statistik.baden-wuerttemberg.de/Statistik-Portal/de_jb27_jahrtab65.asp)

[7] See the list of countries by GDP (nominal).

[8] "Kirchenmitgliederzahlen am 31. Dezember 2010" (http://www.ekd.de/download/Ber_Kirchenmitglieder_2010.pdf) (PDF). ekd.de. . Retrieved 2012-03-10.

[9] Etwa vier Millionen Muslime in Deutschland (http://www.deutsche-islam-konferenz.de/nn_1325560/SubSites/DIK/DE/Magazin/ZahlenDatenFakten/ZahlMLD/zahl-mld-inhalt.html) DIK-Studie

[10] Königlicher Hirschgarten. "Ein paar Worte zu unserem Biergarten in München ... (in German)" (http://www.hirschgarten.de/biergarten-muenchen-koeniglicher-hirschgarten-der-groesste-der-biergaerten-in-muenchen.html). .

[11] Leavenworth, Washington The Bavarian Village (http://www.leavenworth.org/)

External links

- Churches (http://commons.wikimedia.org/wiki/Churches_in_Bavaria)
- Official Tourism Board (http://www.bavaria.by)
- Official government website (http://www.bayern.de)
- Platform with additional information about many cities in Bavaria (http://www.bayregio.de)
- Tradition and Culture (http://www.bavariandancers.com)
- Foreign Trade (http://www.auwi-bayern.de)
- statistics (http://www.statistik.bayern.de/veroeffentlichungen/)
- Geographic data related to Bavaria (http://www.openstreetmap.org/browse/relation/62549) at OpenStreetMap

Germany

Federal Republic of Germany
Bundesrepublik Deutschland

Motto: Einigkeit und Recht und Freiheit(German)
Unity and Justice and Freedom

Anthem: The third stanza of *Lied der Deutschen*
Song of the Germans[1]

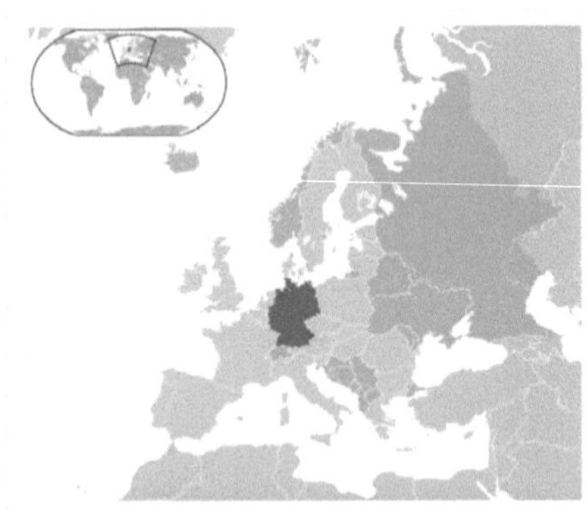

Location of Germany(dark green)
– in Europe(green & dark grey)
– in the European Union(green) — [Legend]

Capital and largest city		Berlin 52°31′N 13°23′E
Official languages		German[1]
Ethnic groups		81% Germans,[2] [3] [4] [5] 7% other Europeans, 4% Turks, 2% Asian, 6% others
Demonym		German
Government		Federal parliamentary constitutional republic
-	President	Joachim Gauck
-	Chancellor	Angela Merkel
-	President of the Bundestag	Norbert Lammert

-	President of the Bundesrat	Winfried Kretschmann
Legislature		
-	Upper house	Bundesrat
-	Lower house	Bundesag
Formation		
-	Holy Roman Empire	2 February 962
-	Unification	18 January 1871
-	Federal Republic	23 May 1949
-	Reunification	3 October 1990
Area		
-	Total	357,021 km^2 (63rd) 137,847 sq mi
-	Water (%)	2.416
Population		
-	2010 estimate	81,799,600[6] (16th)
-	Density	229/km^2 (57th) 593/sq mi
GDP (PPP)		2012 estimate
-	Total	$3.194 trillion[7] (5th)
-	Per capita	$39,059[7] (17th)
GDP (nominal)		2012 estimate
-	Total	$3.367 trillion[7] (4th)
-	Per capita	$41,168[7] (19th)
Gini (2006)		27 (low
HDI (2011)		▲ 0.905[8] (very high / 9th)
Currency		Euro (€) (EUR)
Time zone		CET (UTC+1)
-	Summer (DST)	CEST (UTC+2)
Drives on the		right
Calling code		49
ISO 3166 code		DE
Internet TLD		.de [2]
1.	Danish, Low German, Sorbian, Romany and Frisian are officially recognised by the ECRML.	
2.	The .eu domain is also used, as it is shared with other European Union member states.	

Germany (i/ˈdʒɜrməni/; German: *Deutschland*), officially the **Federal Republic of Germany** (German: *Bundesrepublik Deutschland*, pronounced [ˈbʊndəsʁepuˌbliːk ˈdɔʏtʃlant] (listen)),[9] is a federal parliamentary republic in west-central Europe. The country consists of 16 states, and its capital and largest city is Berlin. Germany covers an area of **unknown operator: u','** square kilometres (**unknown operator: u'strong'** sq mi) and has a largely temperate seasonal climate. With 81.8 million inhabitants, it is the most populous member state in the European Union. Germany is one of the major political and economic powers of the European continent and a historic leader in many theoretical and technical fields.

A region named Germania, inhabited by several Germanic peoples, was documented before AD 100. During the Migration Period, the Germanic tribes expanded southward and established successor kingdoms throughout much of Europe. Beginning in the 10th century, German territories formed a central part of the Holy Roman Empire.[10] During the 16th century, northern German regions became the centre of the Protestant Reformation while southern and western parts remained dominated by Roman Catholic denominations, with the two factions clashing in the Thirty Years' War, marking the beginning of the Catholic–Protestant divide that has characterized German society ever since.[11] Occupied during the Napoleonic Wars, the rise of Pan-Germanism inside the German Confederation resulted in the unification of most of the German states in 1871 into the German Empire, which was Prussian dominated.

After the German Revolution of 1918–1919 and the subsequent military surrender in World War I, the Empire was replaced by the parliamentary Weimar Republic in 1918, and some of its territory partitioned in the Treaty of Versailles. Despite its lead in many scientific and artistic fields at this time, amidst the Great Depression, the Third Reich was established in 1933. The latter period was marked by fascism and World War II. After 1945, Germany was divided by allied occupation, and evolved into two states, East Germany and West Germany. In 1990, the country was reunified.

Germany was a founding member of the European Community in 1957, which became the EU in 1993. It is part of the Schengen Area and since 1999 a member of the euro area. Germany is a great power and member of the United Nations, NATO, the G8, the G20, the OECD and the Council of Europe, and took a non-permanent seat on the UN Security Council for the 2011–2012 term.

Germany has the world's fourth largest economy by nominal GDP and the fifth largest by purchasing power parity. Germany is the second largest exporter and third largest importer of goods. The country has developed a very high standard of living and features a comprehensive system of social security; the country has the world's oldest universal health care system. Germany has been the home of many influential philosophers, music composers, scientists and inventors, and is known for its cultural and political history.

Etymology

The English word *Germany* derives from the Latin Germania, which came into use after Julius Caesar adopted it for the peoples east of the Rhine.[12] The German term *Deutschland* (originally *diutisciu land*, "the German lands") is derived from *deutsch*, descended from Old High German *diutisc* "popular" (i.e. belonging to the *diot* or *diota* "people"), originally used to distinguish the language of the common people from Latin and its Romance descendants. This in turn descends from Proto-Germanic **þiudiskaz* "popular" (see also the Latinised form Theodiscus), derived from **þeudō*, descended from Proto-Indo-European **tewtéh₂-* "people".[13]

History

Germanic tribes and Frankish Empire

Map of the Germania and the Roman Empire

The Germanic tribes are thought to date from the Nordic Bronze Age or the Pre-Roman Iron Age. From southern Scandinavia and north Germany, they expanded south, east and west from the 1st century BC, coming into contact with the Celtic tribes of Gaul as well as Iranian, Baltic, and Slavic tribes in Central and Eastern Europe.[14] Under Augustus, the Roman General Publius Quinctilius Varus began to invade Germania (an area extending roughly from the Rhine to the Ural Mountains). In AD 9, three Roman legions led by Varus were defeated by the Cheruscan leader Arminius. By AD 100, when Tacitus wrote *Germania*, Germanic tribes had settled along the Rhine and the Danube (the Limes Germanicus), occupying most of the area of modern Germany; Austria, southern Bavaria and the western Rhineland, however, were Roman provinces.[15]

In the 3rd century a number of large West Germanic tribes emerged: Alemanni, Franks, Chatti, Saxons, Frisii, Sicambri, and Thuringii. Around 260, the Germanic peoples broke into Roman-controlled lands.[16] After the invasion of the Huns in 375, and with the decline of Rome from 395, Germanic tribes moved further south-west. Simultaneously several large tribes formed in what is now Germany and displaced the smaller Germanic tribes. Large areas (known since the Merovingian period as Austrasia) were occupied by the Franks, and Northern Germany was ruled by the Saxons and Slavs.[15]

Holy Roman Empire

Martin Luther initiated the Protestant Reformation.

On 25 December 800, the Frankish king Charlemagne was crowned emperor and founded the Carolingian Empire, which was divided in 843.[17] The Holy Roman Empire resulted from the eastern portion of this division. Its territory stretched from the Eider River in the north to the Mediterranean coast in the south.[17] Under the reign of the Ottonian emperors (919–1024), several major duchies were consolidated, and the German king was crowned Holy Roman Emperor of these regions in 962. The Holy Roman Empire absorbed northern Italy and Burgundy under the reign of the Salian emperors (1024–1125), although the emperors lost power through the Investiture Controversy.

Under the Hohenstaufen emperors (1138–1254), the German princes increased their influence further south and east into territories inhabited by Slavs, preceding German settlement in these areas and further east *(Ostsiedlung)*. Northern German towns grew prosperous as members of the Hanseatic League.[18] Starting with the Great Famine in 1315, then the Black Death of

1348–50, the population of Germany plummeted.[19] The edict of the Golden Bull in 1356 provided the basic constitution of the empire and codified the election of the emperor by seven prince-electors who ruled some of the most powerful principalities and archbishoprics.[20]

Martin Luther publicised The Ninety-Five Theses in 1517 in Wittenberg, challenging the Roman Catholic Church and initiating the Protestant Reformation. A separate Lutheran church became the official religion in many German states after 1530. Religious conflict led to the Thirty Years' War (1618–1648), which devastated German lands.[21] The population of the German states was reduced by about 30%.[22] The Peace of Westphalia (1648) ended religious warfare among the German states, but the empire was *de facto* divided into numerous independent principalities. In the 18th century, the Holy Roman Empire consisted of approximately 1,800 such territories.[23] From 1740 onwards, dualism between the Austrian Habsburg Monarchy and the Kingdom of Prussia dominated German history. In 1806, the *Imperium* was overrun and dissolved as a result of the Napoleonic Wars.[24]

German Confederation and Empire

Foundation of the German Empire in Versailles, 1871. Bismarck is at the centre in a white uniform.

Following the fall of Napoleon, the Congress of Vienna convened in 1814 and founded the German Confederation (Deutscher Bund), a loose league of 39 sovereign states. Disagreement with restoration politics partly led to the rise of liberal movements, followed by new measures of repression by Austrian statesman Metternich. The *Zollverein*, a tariff union, furthered economic unity in the German states.[25] National and liberal ideals of the French Revolution gained increasing support among many, especially young, Germans. In the light of a series of revolutionary movements in Europe, which established a republic in France, intellectuals and commoners started the Revolutions of 1848 in the German states. King Frederick William IV of Prussia was offered the title of Emperor, but with a loss of power; he rejected the crown and the proposed constitution, leading to a temporary setback for the movement.[26]

Imperial Germany (1871–1918), with the dominant Kingdom of Prussia in blue

Conflict between King William I of Prussia and the increasingly liberal parliament erupted over military reforms in 1862, and the king appointed Otto von Bismarck the new Minister President of Prussia. Bismarck successfully waged war on Denmark in 1864. Prussian victory in the Austro-Prussian War of 1866 enabled him to create the North German Confederation (Norddeutscher Bund) and to exclude Austria, formerly the leading German state, from the federation's affairs. After the French defeat in the Franco-Prussian War, the German Empire was proclaimed in 1871 in Versailles, uniting all scattered parts of Germany except Austria (*Kleindeutschland*, or "Lesser Germany"). With almost two-thirds of its territory and population, Prussia was the dominating constituent of the new state; the Hohenzollern King of Prussia ruled as its concurrent Emperor, and Berlin became its capital.[26] In the *Gründerzeit* period following the unification of Germany, Bismarck's foreign policy as Chancellor of Germany under Emperor William I secured Germany's position as a great nation by forging alliances, isolating France by diplomatic means, and avoiding war. Under Wilhelm II, however, Germany, like other European powers, took an imperialistic course leading to friction with neighbouring countries. As a result of the Berlin Conference in 1884 Germany claimed several colonies including German East Africa, German South-West Africa, Togo, and Cameroon.[27] Most alliances in which Germany had previously been involved were not renewed, and new alliances excluded the country.[28]

The assassination of Austria's crown prince on 28 June 1914 triggered World War I. Germany, as part of the Central Powers, suffered defeat against the Allies in one of the bloodiest conflicts of all time. An estimated two million German soldiers died in World War I.[29] The German Revolution broke out in November 1918, and Emperor Wilhelm II and all German ruling princes abdicated. An armistice ended the war on 11 November, and Germany was forced to sign the Treaty of Versailles in June 1919. The treaty was perceived in Germany as a humiliating continuation of the war, and is often cited as an influence in the rise of Nazism.[30]

Weimar Republic and Nazi Regime

Adolf Hitler, chancellor and president[1] 1933–19451: office formally vacant from August 1934; Hitler styled himself *"Führer und Reichskanzler"*.[31]

At the beginning of the German Revolution in November 1918, Germany was declared a republic. However, the struggle for power continued, with radical-left Communists seizing power in Bavaria. The revolution came to an end on 11 August 1919, when the democratic Weimar Constitution was signed by President Friedrich Ebert.[32] Suffering from the Great Depression of 1929, the harsh peace conditions dictated by the Treaty of Versailles, and a long succession of unstable governments, Germans increasingly lacked identification with the government. This was exacerbated by a widespread right-wing *Dolchstoßlegende*, or stab-in-the-back legend, which argued that Germany had lost World War I because of those who wanted to overthrow the government. The Weimar government was accused of betraying Germany by signing the Versailles Treaty. By 1932, the German Communist Party and the Nazi Party controlled the majority of Parliament, fuelled by discontent with the Weimar government. After a series of unsuccessful cabinets, President Paul von Hindenburg appointed Adolf Hitler as Chancellor of Germany on 30 January 1933.[33] On 27 February 1933 the Reichstag building went up in flames, and a consequent emergency decree abrogated basic citizens' rights. An enabling act passed in parliament gave Hitler unrestricted legislative power. Only the Social Democratic Party voted against it, while Communist MPs had already been imprisoned.[34] [35] Using his powers to crush any actual or potential resistance, Hitler established a centralised totalitarian state within months. Industry was revitalised with a focus on military rearmament.[36]

In 1935, Germany reacquired control of the Saar and in 1936 military control of the Rhineland, both of which had been lost in the Treaty of Versailles.[37] In 1938, Austria was annexed, and in 1939, Czechoslovakia was brought under German control. The invasion of Poland was prepared through the Molotov–Ribbentrop pact and Operation Himmler. On 1 September 1939 the German Wehrmacht launched a blitzkrieg on Poland, which was swiftly occupied by Germany and by the Soviet Red Army. The UK and France declared war on Germany, marking the beginning of World War II.[38] As the war progressed, Germany and its allies quickly gained control of most of continental Europe and North Africa, though plans to force the United Kingdom to an armistice or surrender failed. On 22 June 1941, Germany broke the Molotov–Ribbentrop pact and invaded the Soviet Union. Japan's attack on Pearl Harbor led Germany to declare war on the United States. The Battle of Stalingrad forced the German army to retreat on the Eastern front.[38] In September 1943, Germany's ally Italy surrendered, and German troops were forced to defend an additional front in Italy. D-Day opened a Western front, as Allied forces advanced towards German territory. On 8 May 1945, the German armed forces surrendered after the Red Army occupied Berlin.[39]

In what later became known as The Holocaust, the Third Reich regime had enacted policies directly subjugating many dissidents and minorities. Millions of people were murdered by the Nazis during the Holocaust, including several million Jews, Romani people, Jehovah's Witnesses, Slavic people, Soviet POWs, people with mental and/or physical disabilities, homosexuals, and members of the political and religious opposition.[40] World War II was responsible for more than 40 million dead in Europe.[41] The Nuremberg trials of Nazi war criminals were held after World War II.[42] The war casualties for Germany are estimated at 5.3 million German soldiers[43] millions of German civilians;[44] [45] [46] [47] [48] and losing the war resulted in large territorial losses; the expulsion of about 15 million ethnic Germans from former eastern territories of Germany and other formerly occupied European countries; mass rape of German women;[49] and the destruction of numerous major cities.

Berlin in ruins after World War II

East and West Germany

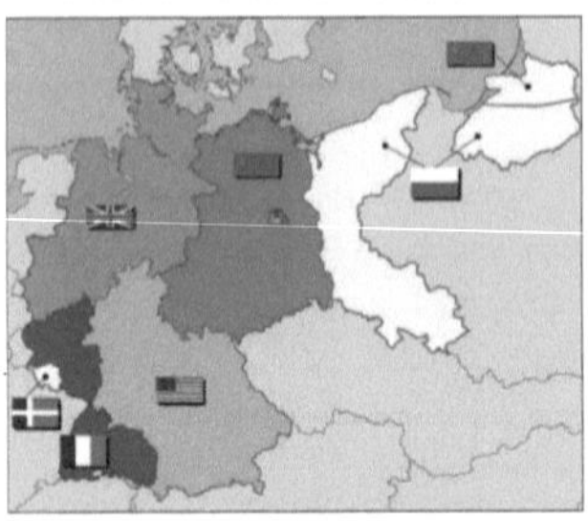

Occupation zones in Germany, 1947. The territories east of the Oder-Neisse line, under Polish and Soviet *de jure* administration and *de facto* annexation, are shown as white as is the detached Saar protectorate.

After the surrender of Germany, the remaining German territory and Berlin were partitioned by the Allies into four military occupation zones. Together, these zones accepted more than 6.5 million of the ethnic Germans expelled from eastern areas.[50] The western sectors, controlled by France, the United Kingdom, and the United States, were merged on 23 May 1949 to form the *Federal Republic of Germany* (*Bundesrepublik Deutschland*); on 7 October 1949, the Soviet Zone became the *German Democratic Republic* (*Deutsche Demokratische Republik*, or DDR). They were informally known as "West Germany" and "East Germany". East Germany selected East Berlin as its capital, while West Germany chose Bonn as a provisional capital, to emphasise its stance that the two-state solution was an artificial and temporary *status quo*.[51]

West Germany, established as a federal parliamentary republic with a "social market economy", was allied with the United States, the UK and France. The country enjoyed prolonged economic growth beginning in the early 1950s (*Wirtschaftswunder*). West Germany joined NATO in 1955 and was a founding member of the European Economic Community in 1957. East Germany was an Eastern Bloc state under political and military control by the USSR via the latter's occupation forces and the Warsaw Pact. Though East Germany claimed to be a democracy, political power was exercised solely by leading members (*Politbüro*) of the communist-controlled Socialist Unity Party of Germany (SED), supported by the Stasi, an immense secret service,[52] and a variety of sub-organisations controlling every aspect of society. A Soviet-style command economy was set up; the GDR later became a Comecon state.[53] While East German propaganda was based on the benefits of the GDR's social programmes and the alleged constant threat of a West German invasion, many of her citizens looked to the West for freedom and prosperity.[54] The Berlin Wall, built in 1961 to stop East Germans from escaping to West Germany, became a symbol of the Cold War,[26] hence its fall in 1989, following democratic reforms in Poland and Hungary, became a symbol of the Fall of Communism, German Reunification and *Die Wende*.

Tensions between East and West Germany were reduced in the early 1970s by Chancellor Willy Brandt's *Ostpolitik.* In summer 1989, Hungary decided to dismantle the Iron Curtain and open the borders, causing the emigration of thousands of East Germans to West Germany via Hungary. This had devastating effects on the GDR, where regular mass demonstrations received increasing support. The East German authorities unexpectedly eased the border restrictions, allowing East German citizens to travel to the West; originally intended to help retain East Germany as a state, the opening of the border actually led to an acceleration of the *Wende* reform process. This culminated in the *Two Plus Four Treaty* a year later on 12 September 1990, under which the four occupying powers renounced their rights under the Instrument of Surrender, and Germany regained full sovereignty. This permitted German reunification on 3 October 1990, with the accession of the five re-established states of the former GDR (new states or "neue Länder").[26]

The Berlin Wall in front of the Brandenburg Gate shortly before its fall in 1989

German reunification and the EU

Based on the Berlin/Bonn Act, adopted on 10 March 1994, Berlin once again became the capital of the reunified Germany, while Bonn obtained the unique status of a *Bundesstadt* (federal city) retaining some federal ministries.[55] The relocation of the government was completed in 1999.[56] Since reunification, Germany has taken a more active role in the European Union and NATO. Germany sent a peacekeeping force to secure stability in the Balkans and sent a force of German troops to Afghanistan as part of a NATO effort to provide security in that country after the ousting of the Taliban.[57] These deployments were controversial since, after the war, Germany was bound by domestic law only to deploy troops for defence roles.[58] In 2005, Angela Merkel became the first female Chancellor of Germany as the leader of a grand coalition.[26]

Geography

Germany is in Western and Central Europe, with Denmark bordering to the north, Poland and the Czech Republic to the east, Austria and Switzerland to the south, France and Luxembourg to the southwest, and Belgium and the Netherlands to the northwest. It lies mostly between latitudes 47° and 55° N (the tip of Sylt is just north of 55°), and longitudes 5° and 16° E. The territory covers 357021 km^2 (137847 sq mi), consisting of 349223 km^2 (134836 sq mi) of land and 7798 km^2 (3011 sq mi) of water. It is the seventh largest country by area in Europe and the 62nd largest in the world.[1]

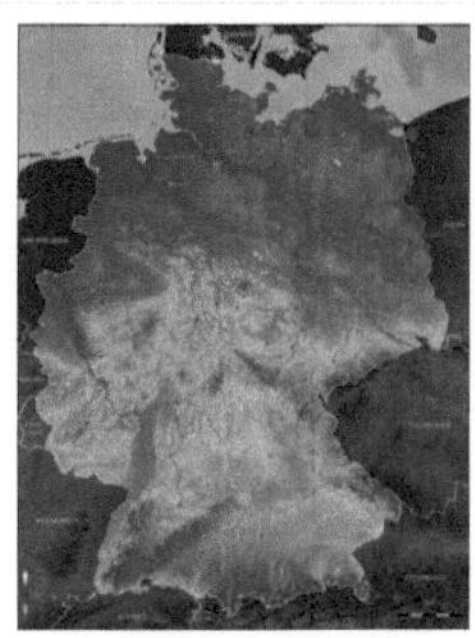

Topographic map

Elevation ranges from the mountains of the Alps (highest point: the Zugspitze at 2962 metres / 9718 feet) in the south to the shores of the North Sea (*Nordsee*) in the northwest and the Baltic Sea (*Ostsee*) in the northeast. The forested uplands of central Germany and the lowlands of northern Germany (lowest point: Wilstermarsch at 3.54 metres / 11.6 feet below sea level) are traversed by such major rivers as the Rhine, Danube and Elbe. Glaciers are found in the Alpine region, but are experiencing deglaciation. Significant natural resources are iron ore, coal, potash, timber, lignite, uranium, copper, natural gas, salt, nickel, arable land and water.[1]

Climate

Most of Germany has a temperate seasonal climate in which humid westerly winds predominate. The climate is moderated by the North Atlantic Drift, the northern extension of the Gulf Stream. This warmer water affects the areas bordering the North Sea; consequently in the northwest and the north the climate is oceanic. Rainfall occurs year-round, especially in the summer. Winters are mild and summers tend to be cool, though temperatures can exceed 30 °C (**unknown operator: u'strong'** °F).[59]

The east has a more continental climate; winters can be very cold and summers very warm, and long dry periods are frequent. Central and southern Germany are transition regions which vary from moderately oceanic to continental. In addition to the maritime and continental climates that predominate over most of the country, the Alpine regions in the extreme south and, to a lesser degree, some areas of the Central German Uplands have a mountain climate, characterised by lower temperatures and greater precipitation.[59]

Biodiversity

The eagle is a protected bird of prey and the national heraldic animal.

The territory of Germany can be subdivided into two ecoregions: European-Mediterranean montane mixed forests and Northeast-Atlantic shelf marine.[60] As of 2008 the majority of Germany is covered by either arable land (34%) or forest and woodland (30.1%); only 13.4% of the area consists of permanent pastures, 11.8% is covered by settlements and streets.[61]

Plants and animals are those generally common to middle Europe. Beeches, oaks, and other deciduous trees constitute one-third of the forests; conifers are increasing as a result of reforestation. Spruce and fir trees predominate in the upper mountains, while pine and larch are found in sandy soil. There are many species of ferns, flowers, fungi, and mosses. Wild animals include deer, wild boar, mouflon, fox, badger, hare, and small numbers of beavers.[62]

The national parks in Germany include the Wadden Sea National Parks, the Jasmund National Park, the Vorpommern Lagoon Area National Park, the Müritz National Park, the Lower Oder Valley National Park, the Harz National Park, the Saxon Switzerland National Park and the Bavarian Forest National Park. More than 400 registered zoos and animal parks operate in Germany, which is believed to be the largest number in any country.[63] The Zoologische Garten Berlin is the oldest zoo in Germany and presents the most comprehensive collection of species in the world.[64]

Politics

Germany is a federal, parliamentary, representative democratic republic. The German political system operates under a framework laid out in the 1949 constitutional document known as the *Grundgesetz* (Basic Law). Amendments generally require a two-thirds majority of both chambers of parliament; the fundamental principles of the constitution, as expressed in the articles guaranteeing human dignity, the separation of powers, the federal structure, and the rule of law are valid in perpetuity.[65]

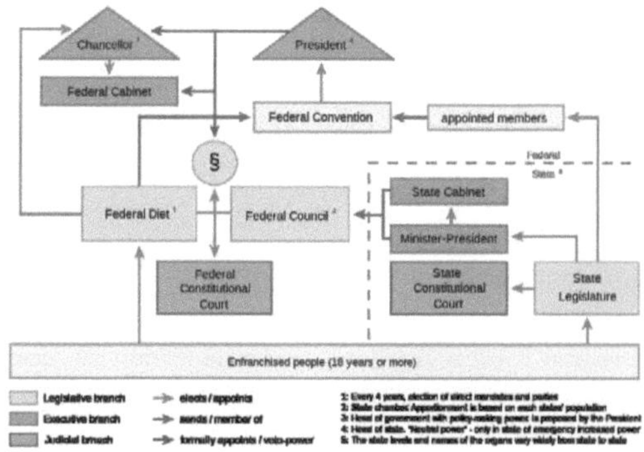

Political system of Germany

The president is the head of state and invested primarily with representative responsibilities and powers. He is elected by the *Bundesversammlung* (federal convention), an institution consisting of the members of the *Bundestag* and an equal number of state delegates. The second-highest official in the German order of precedence is the *Bundestagspräsident* (President of the *Bundestag*), who is elected by the *Bundestag* and responsible for overseeing the daily sessions of the body. The third-highest official and the head of government is the Chancellor, who is appointed by the *Bundespräsident* after being elected by the *Bundestag*.[26]

The Reichstag building in Berlin is the site of the German parliament (Bundestag).

Joachim Gauck, President of Germany

The chancellor, currently Angela Merkel, is the head of government and exercises executive power, similar to the role of a Prime Minister in other parliamentary democracies. Federal legislative power is vested in the parliament consisting of the *Bundestag* (Federal Diet) and *Bundesrat* (Federal Council), which together form the legislative body. The *Bundestag* is elected through direct elections, by proportional representation (mixed-member).[1] The members of the *Bundesrat* represent the governments of the sixteen federated states and are members of the state cabinets.[26]

Since 1949, the party system has been dominated by the Christian Democratic Union and the Social Democratic Party of Germany. So far every chancellor has been a member of one of these parties. However, the smaller liberal Free Democratic Party (which has had members in the *Bundestag* since 1949) and the Alliance '90/The Greens (which has had seats in parliament since 1983) have also played important roles.[66]

Germany has a civil law system based on Roman law with some references to Germanic law. The *Bundesverfassungsgericht* (Federal Constitutional Court) is the German Supreme Court responsible for constitutional matters, with power of judicial review.[26] [67] Germany's supreme court system, called *Oberste Gerichtshöfe des Bundes*, is specialised: for civil and criminal cases, the highest court of appeal is the inquisitorial Federal Court of Justice, and for other affairs the courts are the Federal Labour Court, the Federal Social Court, the Federal Finance Court and the Federal Administrative Court. The *Völkerstrafgesetzbuch* regulates the consequences of crimes against humanity, genocide and war crimes, and gives German courts universal jurisdiction in some circumstances.[68] Criminal and private laws are codified on the national level in the *Strafgesetzbuch* and the *Bürgerliches Gesetzbuch* respectively. The German penal system is aimed towards rehabilitation of the criminal and the protection of the general public.[69] Except for petty crimes, which are tried before a single professional judge, and serious political crimes, all charges are tried before mixed tribunals on which lay judges (*Schöffen*) sit side by side with professional judges.[70] [71]

Constituent states

Germany comprises sixteen states which are collectively referred to as *Länder*.[72] Each state has its own state constitution[73] and is largely autonomous in regard to its internal organisation. Due to differences in size and population the subdivisions of these states vary, especially as between city states (*Stadtstaaten*) and states with larger territories (*Flächenländer*). For regional administrative purposes five states, namely Baden-Württemberg, Bavaria, Hesse, North Rhine-Westphalia and Saxony, consist of a total of 22 Government Districts (*Regierungsbezirke*). As of 2009 Germany is divided into 403 districts (*Kreise*) at a municipal level; these consist of 301 rural districts and 102 urban districts.[74]

Lower Saxony

Bremen

Hamburg

Mecklenburg-Vorpommern

Saxony-Anhalt

Saxony

Brandenburg

Berlin

Thuringia

Hesse

North Rhine-Westphalia

Rhineland-Palatinate

Bavaria

Baden-Württemberg

Saarland

Schleswig-Holstein

State	Capital	Area (km²)	Population
Baden-Württemberg	Stuttgart	35,752	10,753,880
Bavaria	Munich	70,549	12,538,696
Berlin	Berlin	892	3,460,725
Brandenburg	Potsdam	29,477	2,503,273
Bremen	Bremen	404	660,999
Hamburg	Hamburg	755	1,786,448
Hesse	Wiesbaden	21,115	6,067,021
Mecklenburg-Vorpommern	Schwerin	23,174	1,642,327
Lower Saxony	Hanover	47,618	7,918,293
North Rhine-Westphalia	Düsseldorf	34,043	17,845,154
Rhineland-Palatinate	Mainz	19,847	4,003,745
Saarland	Saarbrücken	2,569	1,017,567
Saxony	Dresden	18,416	4,149,477
Saxony-Anhalt	Magdeburg	20,445	2,335,006
Schleswig-Holstein	Kiel	15,763	2,834,259
Thuringia	Erfurt	16,172	2,235,025

Foreign relations

Chancellor Angela Merkel hosting the G8 summit in Heiligendamm

Germany has a network of 229 diplomatic missions abroad[75] and maintains relations with more than 190 countries.[76] As of 2011 it is the largest contributor to the budget of the European Union (providing 20%)[77] and the third largest contributor to the UN (providing 8%).[78] Germany is a member of NATO, the Organisation of Economic Co-operation and Development (OECD), the G8, the G20, the World Bank and the International Monetary Fund (IMF). It has played a leading role in the European Union since its inception and has maintained a strong alliance with France since the end of World War II. Germany seeks to advance the creation of a more unified European political, defence, and security apparatus.[79] [80]

The development policy of the Federal Republic of Germany is an independent area of German foreign policy. It is formulated by the Federal Ministry for Economic Cooperation and Development (BMZ) and carried out by the implementing organisations. The German government sees development policy as a joint responsibility of the international community.[81] It is the world's third biggest aid donor after the United States and France.[82] [83]

During the Cold War, Germany's partition by the Iron Curtain made it a symbol of East-West tensions and a political battleground in Europe. However, Willy Brandt's Ostpolitik was a key factor in the *détente* of the 1970s.[84] In 1999, Chancellor Gerhard Schröder's government defined a new basis for German foreign policy by taking part in the NATO decisions surrounding the Kosovo War and by sending German troops into combat for the first time since World War II.[85] The governments of Germany and the United States are close political allies.[26] The 1948 Marshall Plan and strong cultural ties have crafted a strong bond between the two countries, although Schröder's vocal opposition to the Iraq War suggested the end of Atlanticism and a relative cooling of German-American relations.[86] The two countries are also economically interdependent: 8.8% of German exports are U.S.-bound and 6.6% of German imports originate from the U.S.[87]

Military

Germany's military, the *Bundeswehr*, is organized into *Heer* (Army), *Marine* (Navy), *Luftwaffe* (Air Force), *Bundeswehr Joint Medical Service* and *Streitkräftebasis* (Joint Support Service) branches. As of 2011, military spending was an estimated 1.3% of the country's GDP, which is low in a ranking of all countries; in absolute terms, German military expenditure is the 9th highest in the world.[88] In peacetime, the Bundeswehr is commanded by the Minister of Defence. If Germany went to war, which according to the constitution is allowed only for defensive purposes, the Chancellor would become commander-in-chief of the *Bundeswehr*.[89]

As of March 2012 the Bundeswehr employs 183,000 professional soldiers and 17,000 volunteers.[90] The German government plans to reduce the number of soldiers to 170,000 professionals and up to 15,000 short-term volunteers (*voluntary military service*).[91] Reservists are available to the Armed Forces and participate in defence exercises and deployments abroad.[91] As of April 2011, the German military had about 6,900 troops stationed in foreign countries as part of international peacekeeping forces, including about 4,900 Bundeswehr troops in the NATO-led ISAF force in Afghanistan and Uzbekistan, 1,150 German soldiers in Kosovo, and 300 troops with UNIFIL in Lebanon.[92]

Until 2011, military service was compulsory for men at age 18, and conscripts served six-month tours of duty; conscientious objectors could instead opt for an equal length of *Zivildienst* (civilian service), or a six-year commitment to (voluntary) emergency services like a fire department or the Red Cross. On 1 July 2011 conscription was officially suspended and replaced with a voluntary service.[93] [94] Since 2001 women may serve in all functions of service without restriction, but they are not subject to conscription. There are presently some 17,500 women on active duty and a number of female reservists.[95]

Economy

A Mercedes-Benz car. Germany was the world's leading exporter of goods from 2003 to 2008.[96]

Germany has a social market economy with a highly skilled labour force, a large capital stock, a low level of corruption,[97] and a high level of innovation.[98] It has the largest and most powerful national economy in Europe, the fourth largest by nominal GDP in the world,[99] the fifth largest by PPP,[100] and was the biggest net contributor to the EU budget in 2011.[101] The service sector contributes approximately 71% of the total GDP, industry 28%, and agriculture 1%.[1] The official average national unemployment rate in May 2012 was 6.7%.[102] However, the official average national unemployment rate also includes people with a part-time job that are looking for a full-time job.[103] The unofficial average national unemployment rate in 2011 was 5.7%.[1]

Germany is an advocate of closer European economic and political integration. Its commercial policies are increasingly determined by agreements among European Union (EU) members and by EU legislation. Germany introduced the common European currency, the euro, on 1 January 2002.[104] [105] Its monetary policy is set by the European Central Bank. Two decades after German reunification, standards of living and per capita incomes remain significantly higher in the states of the former West Germany than in the former East.[106] The modernisation and integration of the eastern German economy is a long-term process scheduled to last until the year 2019, with annual transfers from west to east amounting to roughly $80 billion.[107] In January 2009 the German government approved a €50 billion economic stimulus plan to protect several sectors from a downturn and a subsequent rise in unemployment rates.[108]

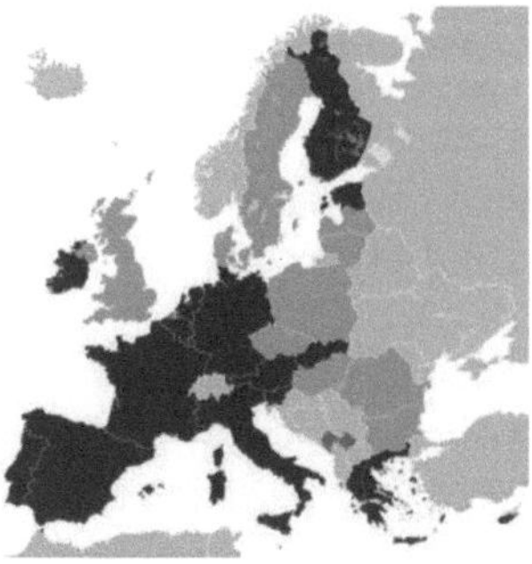

Germany is part of a monetary union, the Eurozone (dark blue), and of the EU single market.

Of the world's 500 largest stock-market-listed companies measured by revenue in 2010, the Fortune Global 500, 37 are headquartered in Germany. 30 Germany-based companies are included in the DAX, the German stock market index. Well-known global brands are Mercedes-Benz, BMW, SAP, Siemens, Volkswagen, Adidas, Audi, Allianz, Porsche, Bayer, Bosch, and Nivea.[109] Germany is recognised for its specialised small and medium enterprises. Around 1,000 of these companies are global market leaders in their segment and are labelled hidden champions.[110]

The list includes the largest companies by turnover in 2009. Unranked are the largest bank and the largest insurance company in 2007:

Rank[111]	Name	Headquarters	Revenue (Mil. €)	Profit (Mil. €)	Employees (World)
1	Volkswagen AG	Wolfsburg	108,897	4,120	329,305
2	Daimler AG	Stuttgart	99,399	3,985	272,382
3	Siemens AG	Munich/Berlin	72,488	3,806	398,200
4	E.ON AG	Düsseldorf	68,731	7,204	87,815
5	Metro AG	Düsseldorf	64,337	825	242,378
6	Deutsche Post AG	Bonn	63,512	1,389	475,100
7	Deutsche Telekom AG	Bonn	62,516	569	241,426
8	BASF SE	Ludwigshafen	57,951	4,065	95,175
9	BMW AG	Munich	56,018	3,126	107,539
10	ThyssenKrupp AG	Essen/Duisburg	51,723	2,102	191,350

Infrastructure

The ICE 3 train at Frankfurt Hauptbahnhof

With its central position in Europe, Germany is a transport hub. This is reflected in its dense and modern transport networks. The motorway (Autobahn) network ranks as the third largest worldwide in length and is known for its lack of a general speed limit.[112] Germany has established a polycentric network of high-speed trains. The InterCityExpress or *ICE* network of the Deutsche Bahn serves major German cities as well as destinations in neighbouring countries with speeds up to 300 kph (186 mph).[113] The largest German airports are Frankfurt Airport and Munich Airport, both hubs of Lufthansa, while Air Berlin has hubs at Berlin Tegel and Düsseldorf. Other major airports include Berlin Schönefeld, Hamburg, Cologne/Bonn and Leipzig/Halle. Both airports in Berlin will be consolidated at a site adjacent to Berlin Schönefeld, which will become Berlin Brandenburg Airport in 2013.[114][115]

As of 2008, Germany was the world's sixth largest consumer of energy,[116] and 60% of its primary energy was imported.[117] Government policy promotes energy conservation and renewable energy. Energy efficiency has been improving since the early 1970s; the government aims to meet the country's electricity demands using 40% renewable sources by 2020 and 100% by 2050.[118] In 2010, energy sources were: oil (33.7%); coal, including lignite (22.9%); natural gas (21.8%); nuclear (10.8%); hydro-electric and wind power (1.5%); and other renewable sources (7.9%).[119] In 2000, the government and the nuclear power industry agreed to phase out all nuclear power plants by 2021.[120] Germany is committed to the Kyoto protocol and several other treaties promoting biodiversity, low emission standards, recycling, and the use of renewable energy, and supports sustainable development at a global level.[121] The German government has initiated wide-ranging emission reduction activities and the country's overall emissions are falling.[122] Nevertheless the country's greenhouse gas emissions were the highest in the EU as of 2010.[123]

Science and technology

Albert Einstein

Germany's achievements in the sciences have been significant, and research and development efforts form an integral part of the economy.[124] The Nobel Prize has been awarded to 103 German laureates.[125] For most of the 20th century, German laureates had more awards than those of any other nation, especially in the sciences (physics, chemistry, and physiology or medicine).[126] [127]

The work of Albert Einstein and Max Planck was crucial to the foundation of modern physics, which Werner Heisenberg and Max Born developed further.[128] They were preceded by such key physicists as Hermann von Helmholtz, Joseph von Fraunhofer and Gabriel Daniel Fahrenheit, among others. Wilhelm Röntgen discovered X-rays and was the first winner of the Nobel Prize in Physics in 1901.[129] Otto Hahn was a pioneer in the fields of radioactivity and radiochemistry and discovered nuclear fission,[130] while Ferdinand Cohn and Robert Koch were founders of microbiology. Numerous mathematicians were born in Germany, including Carl Friedrich Gauss, David Hilbert, Bernhard Riemann, Gottfried Leibniz, Karl Weierstrass, Hermann Weyl and Felix Klein. Research institutions in Germany include the Max Planck Society, the Helmholtz Association and the Fraunhofer Society. The Gottfried Wilhelm Leibniz Prize is granted to ten scientists and academics every year. With a maximum of €2.5 million per award it is one of highest endowed research prizes in the world.[131]

Germany has been the home of many famous inventors and engineers, such as Johannes Gutenberg, credited with the invention of movable type printing in Europe; Hans Geiger, the creator of the Geiger counter; and Konrad Zuse, who built the first fully automatic digital computer.[132] German inventors, engineers and industrialists such as Count Ferdinand von Zeppelin, Otto Lilienthal, Gottlieb Daimler, Rudolf Diesel, Hugo Junkers and Karl Benz helped shape modern automotive and air transportation technology.[133] Aerospace engineer Wernher von Braun developed the first space rocket and later on was a prominent member of NASA and developed the Saturn V Moon rocket, which paved the way for the success of the US Apollo program. Heinrich Rudolf Hertz's work in the domain of electromagnetic radiation was pivotal to the development of modern telecommunication.[134]

Germany is one of the leading countries in developing and using green technologies. Companies specializing in green technology have an estimated turnover of €200 billion. Key sectors of Germany's green technology industry are power generation, sustainable mobility, material efficiency, energy efficiency, waste management and recycling, and sustainable water management.[135]

Demographics

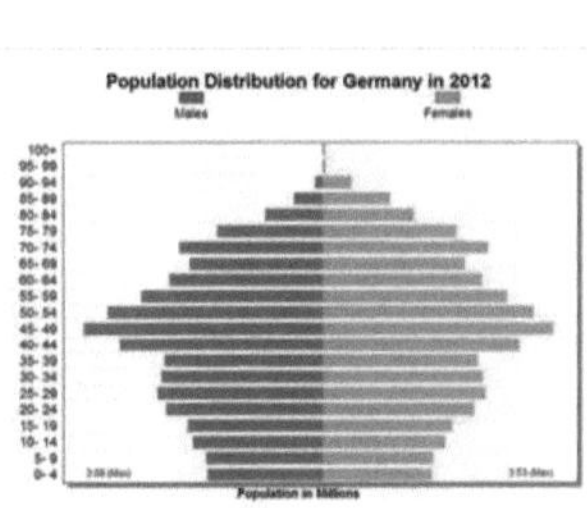

Population pyramid of Germany in 2012 from International Futures

With its estimated population of 81.8 million in January 2010,[6] Germany is the most populous country in the European Union and ranks as the 16th most populous country in the world.[136] Its population density stands at 229.4 inhabitants per square kilometre. The overall life expectancy in Germany at birth is 80.19 years (77.93 years for males and 82.58 years for females).[1] The fertility rate of 1.41 children born per woman (2011 estimates), or 8.33 births per 1000 inhabitants, is one of the lowest in the world.[1] Since the 1970s, Germany's death rate has continuously exceeded its birth rate.[137] The Federal Statistical Office of Germany forecast that the population will shrink to between 65 and 70 million by 2060 (depending on the level of net migration).[138]

Germans by nationality make up 91% of the population of Germany. As of 2010, about seven million foreign citizens were registered in Germany, and 20%[2] of the country's residents, or more than 16 million people, were of foreign or partially foreign descent (including persons descending or partially descending from ethnic German repatriates), 96% of whom lived in the former West Germany or Berlin.[139]

In 2010, 2.3 million families with children under 18 years were living in Germany, in which at least one parent had foreign roots. They represented 29% of the total of 8.1 million families with minor children. Compared with 2005 – the year when the microcensus started to collect detailed information on the population with a migrant background – the proportion of migrant families has risen by 2 percentage points.[140]

Most of the families with a migrant background live in the western part of Germany. In 2010, the proportion of migrant families in all families was 32% in the pre-unification territory of the Federal Republic. This figure was more than double that in the new Länder (including Berlin) where it stood at 15%.[140]

Families with a migrant background more often have three or more minor children in the household than families without a migrant background. In 2010, about 15% of the families with a migrant background contained three or more minor children, as compared with just 9% of the families without a migrant background.[140]

The United Nations Population Fund lists Germany as host to the third-highest number of international migrants worldwide, about 5% or 10 million of all 191 million migrants.[141] As a consequence of restrictions to Germany's formerly rather unrestricted laws on asylum and immigration, the number of immigrants seeking asylum or claiming German ethnicity (mostly from the former Soviet Union) has been declining steadily since 2000.[142] In 2009, 20% of the population had immigrant roots, the highest since 1945.[143] As of 2008, the largest national group was from Turkey (2.5 million), followed by Italy (776,000) and Poland (687,000).[144] About 3 million "Aussiedler"—ethnic Germans, mainly from the former eastern bloc—have resettled in Germany since 1987.[145] Most ethnic minorities (especially those of non-European origin) reside in large urban areas like Berlin, Hamburg, Frankfurt Rhine-Main, Rhine-Ruhr, Rhine-Neckar and Munich. The percentage of non-Germans and immigrants is quite low in rural areas and small towns, especially in the former East German states of Brandenburg, Saxony and Mecklenburg-Vorpommern.

Ethnic makeup as of 2010:

Ethnic Group	**% of Germany's population[146] [147]**	**population**
European	**88.0**	71,935,000
Ethnic German	80.7	65,970,000
Polish	2.0	1,654,000
former Soviet Union (primarily Russian Germans, Russians and Jews)	1.7	1,400,000
European Other (most notably Southern Europeans, Western Europeans and former Yugoslavians)	3.6	3,000,000
Middle Eastern	**5.2**	4,260,000
Turkish	4.0	3,260,000
others (primarily Arabs and Iranians)	1.2	1,000,000
Asian (especially Vietnamese and Thai people)	**2.0**	1,634,000
Afro-German or Black African	**1.0**	817,150
Mixed or unspecified background	**2.0**	1,634,000
Other groups (primarily the Americas)	**1.8**	1,470,000
Total population	**100**	81,715,000

Germany has a number of large cities. The largest conurbation is the Rhine-Ruhr region (11.7 million as of 2008), including Düsseldorf (the capital of North Rhine-Westphalia), Cologne, Bonn, Dortmund, Essen, Duisburg, and Bochum.[148]

Religion

Christianity is the largest religion in Germany, with around 51.5 million adherents (62.8%) in 2008.[149] Relative to the whole population, 30.0% of Germans are Catholics, 29.9% are Protestants belonging to the Evangelical Church in Germany (EKD), and the remaining Christians belong to smaller denominations each with less than 0.5% of the German population.[150] Protestantism is concentrated in the north and east and Roman Catholicism is concentrated in the south and west;[11] 1.6% of the country's overall population declare themselves Orthodox Christians.[149]

The Cologne Cathedral at the Rhine river is a UNESCO World Heritage Site.

The second largest religion is Islam with an estimated 3.8 to 4.3 million adherents (4.6% to 5.2%),[151] followed by Buddhism with 250,000 and Judaism with around 200,000 adherents (0.3%); Hinduism has some 90,000 adherents (0.1%). All other religious communities in Germany have fewer than 50,000 adherents.[152] Of the roughly 4 million Muslims, most are Sunnis and Alevites from Turkey, but there are a small number of Shi'ites and other denominations.[151] German Muslims, a large portion of whom are of Turkish origin, lack full official state recognition of their religious community.[11] Germany has Europe's third largest Jewish population (after France and the United Kingdom).[153] Approximately 50% of the Buddhists in Germany are Asian immigrants.[154]

Germans with no stated religious adherence make up 34.1% of the population and are concentrated in the former East Germany and major metropolitan areas.[150] German reunification in 1990 greatly increased the country's non-religious population, a legacy of the state atheism of the previously Soviet-controlled East. Christian church membership has decreased in recent decades, particularly among Protestants.[11]

Languages

German is the official and predominant spoken language in Germany.[155] It is one of 23 official languages in the European Union, and one of the three working languages of the European Commission. Recognised native minority languages in Germany are Danish, Low German, Sorbian, Romany, and Frisian; they are officially protected by the European Charter for Regional or Minority Languages. The most used immigrant languages are Turkish, Kurdish, Polish, the Balkan languages, and Russian; 67% of German citizens claim to be able to communicate in at least one foreign language and 27% in at least two languages other than their own.[155]

Standard German is a West Germanic language and is closely related to and classified alongside English, Low German, Dutch, and the Frisian languages. To a lesser extent, it is also related to the East (extinct) and North Germanic languages. Most German vocabulary is derived from the Germanic branch of the Indo-European language family.[156] Significant minorities of words are derived from Latin and Greek, with a smaller amount from French and most recently English (known as Denglisch). German is written using the Latin alphabet. German dialects, traditional local varieties traced back to the Germanic tribes, are distinguished from varieties of standard German by their lexicon, phonology, and syntax.[157]

Education

Heidelberg University was established in 1386.

Over 99% of Germans age 15 and above are estimated to be able to read and write.[1] However, a growing number of inhabitants are functionally illiterate.[158] Responsibility for educational oversight in Germany lies primarily with the individual federated states. Since the 1960s, a reform movement attempted to unify secondary education in a *Gesamtschule* (comprehensive school); several West German states later simplified their school system to two or three tiers. A system of apprenticeship called *Duale Ausbildung* ("dual education") allows pupils in vocational training to learn in a company as well as in a state-run vocational school.[159]

Optional kindergarten education is provided for all children between three and six years old, after which school attendance is compulsory for at least nine years. Primary education usually lasts for four years and public schools are not stratified at this stage.[159] In contrast, secondary education includes three traditional types of schools focused on different levels of academic ability: the *Gymnasium* enrols the most gifted children and prepares students for university studies; the *Realschule* for intermediate students lasts six years; the *Hauptschule* prepares pupils for vocational education.[160]

The general entrance requirement for university is Abitur, a qualification normally based on continuous assessment during the last few years at school and final examinations; however there are a number of exceptions, and precise requirements vary, depending on the state, the university and the subject. Germany's universities are recognised internationally; in the Academic Ranking of World Universities (ARWU) for 2008, six of the top 100 universities in the world are in Germany, and 18 of the top 200.[161] Nearly all German universities are public institutions, charging tuition fees of €50–500 per semester for each student.[162]

Health

Germany has the world's oldest universal health care system, dating back to Bismarck's social legislation in 1883.[163] Currently the population is covered by a fairly comprehensive health insurance plan provided by statute. Certain groups of people (lifetime officials, self-employed persons, employees with high income) can opt out of the plan and switch to a private insurance contract. Previously, these groups could also choose to do without insurance, but this option was dropped in 2009.[164] According to the World Health Organization, Germany's health care system was 77% government-funded and 23% privately funded as of 2005.[165] In 2005, Germany spent 11% of its GDP on health care. Germany ranked 20th in the world in life expectancy with 77 years for men and 82 years for women, and it had a very low infant mortality rate (4 per 1,000 live births).[165]

As of 2010, the principal cause of death was cardiovascular disease, at 41%, followed by malignant tumours, at 26%.[166] As of 2008, about 82,000 Germans had been infected with HIV/AIDS and 26,000 had died from the disease (cumulatively, since 1982).[167] According to a 2005 survey, 27% of German adults are smokers.[167]

Culture

Ludwig van Beethoven (1770–1827), composer

From its roots, culture in German states has been shaped by major intellectual and popular currents in Europe, both religious and secular. Historically Germany has been called *Das Land der Dichter und Denker* ("the land of poets and thinkers"),[168] because of the major role its famous writers and philosophers have played in the development of Western thought and culture.

The federated states are in charge of the cultural institutions. There are 240 subsidised theatres, hundreds of symphonic orchestras, thousands of museums and over 25,000 libraries spread in Germany. These cultural opportunities are enjoyed by many: there are over 91 million German museum visits every year; annually, 20 million go to theatres and operas; 3.6 million per year listen to the symphonic orchestras.[169] As of 2012 the UNESCO inscribed 37 properties in Germany on the World Heritage List.[170]

Germany has established a high level of gender equality,[171] promotes disability rights, and is legally and socially tolerant towards homosexuals. Gays and lesbians can legally adopt their partner's biological children, and civil unions have been permitted since 2001.[172] Germany has also changed its attitude towards immigrants; since the mid-1990s, the government and the majority of Germans have begun to acknowledge that controlled immigration should be allowed based on qualification standards.[173] Germany has been named the world's second most valued nation among 50 countries in 2010.[174] A global opinion poll for the BBC revealed that Germany is recognised for having the most positive influence in the world in 2011.[175]

Arts

J.S.Bach	L.v. Beethoven	R. Wagner
Toccata und Fuge	Symphonie 5 c-moll	Die Walküre

Numerous German painters have enjoyed international prestige through their work in diverse artistic styles. Hans Holbein the Younger, Matthias Grünewald, and Albrecht Dürer were important artists of the Renaissance, Caspar David Friedrich of Romanticism, and Max Ernst of Surrealism. Architectural contributions from Germany include the Carolingian and Ottonian styles, which were precursors of Romanesque. The region later became the site of Gothic, Renaissance and Baroque art. Germany was particularly important in the early modern movement, especially through the Bauhaus movement founded by Walter Gropius. Ludwig Mies van der Rohe became one of the world's most renowned architects in the second half of the 20th century. He conceived of the glass façade skyscraper.[176]

German music includes works by some of the world's most well-known classical music composers, including Ludwig van Beethoven, Johann Sebastian Bach, Wolfgang Amadeus Mozart, Johannes Brahms, Richard Wagner and Richard Strauss. Germany is the largest music market in Europe, and third largest in the world.[177]

Literature and philosophy

The Brothers Grimm

German literature can be traced back to the Middle Ages and the works of writers such as Walther von der Vogelweide and Wolfram von Eschenbach. Well-known German authors include Johann Wolfgang von Goethe, Friedrich Schiller, Gotthold Ephraim Lessing and Theodor Fontane. The collections of folk tales published by the Brothers Grimm popularised German folklore on an international level. Influential authors of the 20th century include Gerhart Hauptmann, Thomas Mann, Hermann Hesse, Heinrich Böll and Günter Grass.[178] German-speaking book publishers produce some 700 million books every year, with about 80,000 titles, nearly 60,000 of them new. Germany comes third in quantity of books published, after the English-speaking book market and the People's Republic of China.[179] The Frankfurt Book Fair is the most important in the world for international deals and trading, with a tradition spanning over 500 years.[180]

German philosophy is historically significant. Gottfried Leibniz's contributions to rationalism; the establishment of classical German idealism by Immanuel Kant, Johann Gottlieb Fichte, Georg Wilhelm Friedrich Hegel and Friedrich Wilhelm Joseph Schelling; Arthur Schopenhauer's composition of metaphysical pessimism; the formulation of communist theory by Karl Marx and Friedrich Engels; Friedrich Nietzsche's development of perspectivism; Gottlob Frege's contributions to the dawn of analytic philosophy; Martin Heidegger's works on Being; and the development of the Frankfurt school by Max Horkheimer, Theodor Adorno, Herbert Marcuse and Jürgen Habermas have been particularly influential. In the 21st century Germany has contributed to the development of contemporary analytic philosophy in continental Europe, along with France, Austria, Switzerland and the Scandinavian countries.[181]

Media

German cinema dates back to the earliest years of the medium with the work of Max Skladanowsky, which was particularly influential with German expressionists such as Robert Wiene and Friedrich Wilhelm Murnau. Director Fritz Lang's *Metropolis* (1927) is referred to as the first modern science-fiction film. In 1930 the Austrian-American Josef von Sternberg directed *The Blue Angel*, the first major German sound film.[182] During the 1970s and 1980s, New German Cinema directors such as Volker Schlöndorff, Werner Herzog, Wim Wenders, and Rainer Werner Fassbinder put West German cinema on the international stage.[183] The annual European Film Awards ceremony is held every other year in Berlin, home of the European Film Academy (EFA); the Berlin Film Festival, held annually since 1951, is one of the world's foremost film festivals.[184]

More recently, films such as *Good Bye, Lenin!* (2003), *Gegen die Wand (Head-On)* (2004), *Der Untergang (Downfall)* (2004), and *The Baader Meinhof Complex* (2008) have had international success. The Academy Award for Best Foreign Language Film went to the German production *Die Blechtrommel (The Tin Drum)* in 1979, to *Nowhere in Africa* in 2002, and to *Das Leben der Anderen (The Lives of Others)* in 2007.[185] Germany's television market is the largest in Europe, with some 34 million TV households. Around 90% of German households have cable or satellite TV, with a variety of free-to-view public and commercial channels.[186]

Cuisine

A *Schwarzwälder Kirschtorte* (literally, "Black Forest cherry torte")

German cuisine varies from region to region. The southern regions of Bavaria and Swabia, for instance, share a culinary culture with Switzerland and Austria. In all regions, meat is often eaten in sausage form.[187] Organic food has gained a market share of ca. 2%, and is expected to increase further.[188] Although wine is becoming more popular in many parts of Germany, the national alcoholic drink is beer. German beer consumption per person is declining, but at 121.4 litres in 2009 it is still among the highest in the world.[189] The Michelin guide has awarded nine restaurants in Germany three stars, the highest designation, while 15 more received two stars.[190] German restaurants have become the world's second-most decorated after France.[191]

Sports

Signal Iduna Park, the stadium of Bundesliga club Borussia Dortmund, is the biggest stadium in Germany.

Twenty-seven million Germans are members of a sports club and an additional twelve million pursue sports individually.[192] Association football is the most popular sport. With more than 6.3 million official members, the German Football Association (*Deutscher Fußball-Bund*) is the largest sports organisation of its kind worldwide.[192] The Bundesliga, the top league of German football, is the most popular sports league in Germany and attracts the second highest average attendance of any professional sports league in the world.

The German national football team won the FIFA World Cup in 1954, 1974 and 1990 and the UEFA European Football Championship in 1972, 1980 and 1996. Germany hosted the FIFA World Cup in 1974 and 2006 and the UEFA European Football Championship in 1988. Among the most well-known footballers are Franz Beckenbauer, Gerd Müller, Jürgen Klinsmann, Lothar Matthäus, and Oliver Kahn. Other popular spectator sports include handball, volleyball, basketball, ice hockey, and tennis.[192]

Germany is one of the leading motor sports countries in the world. Constructors like BMW and Mercedes are prominent manufacturers in motor sport. Additionally, Porsche has won the 24 Hours of Le Mans, an annual endurance race held in France, 16 times, and Audi has won it 11 times. Formula One driver Michael Schumacher has set many motor sport records during his career, having won more Formula One World Drivers' Championships and more Formula One races than any other driver; he is one of the highest paid sportsmen in history.[193]

Historically, German sportsmen have been successful contenders in the Olympic Games, ranking third in an all-time Olympic Games medal count, combining East and West German medals. In the 2008 Summer Olympics, Germany finished fifth in the medal count,[194] while in the 2006 Winter Olympics they finished first.[195] Germany has hosted the Summer Olympic Games twice, in Berlin in 1936 and in Munich in 1972. The Winter Olympic Games took place in Germany once in 1936 in the twin towns of Garmisch and Partenkirchen.

See also

- Index of Germany-related articles
- Outline of Germany

References

[1] "World Factbook" (https://www.cia.gov/library/publications/the-world-factbook/geos/gm.html). CIA. . Retrieved 26 March 2011.

[2] Bevölkerung und Erwerbstätigkeit: Bevölkerung mit Migrationshintergrund – Ergebnisse des Mikrozensus 2010 (https://www.destatis.de/DE/Publikationen/Thematisch/Bevoelkerung/MigrationIntegration/Migrationshintergrund2010220107004.pdf?__blob=publicationFile), p. 64 statistics

[3] Germans without any migrant background (https://www.destatis.de/DE/PresseService/Presse/Pressemitteilungen/2010/01/PD10_033_122.html)

[4] "Press releases - For the first time more than 16 million people with migration background in Germany" (https://www.destatis.de/EN/PressServices/Press/pr/2010/07/PE10_248_122.html). Federal Statistical Office (Destatis) - Destatis.de. 2010-07-14. . Retrieved 2012-11-04.

[5] "Pressemitteilungen - Ein Fünftel der Bevölkerung in Deutschland hatte 2010 einen Migrationshintergrund" (https://www.destatis.de/DE/PresseService/Presse/Pressemitteilungen/2011/09/PD11_355_122.html) (in (German)). Statistisches Bundesamt (Destatis) - Destatis.de. 2011-09-26. . Retrieved 2012-11-04.

[6] *Key Figures on Europe* (http://epp.eurostat.ec.europa.eu/cache/ITY_OFFPUB/KS-EI-11-001/EN/KS-EI-11-001-EN.PDF). Belgium: European Union. 2011. p. 37. doi:10.2785/623. ISBN 978-92-79-18441-3. .

[7] "Germany" (http://www.imf.org/external/pubs/ft/weo/2012/02/weodata/weorept.aspx?pr.x=59&pr.y=9&sy=2012&ey=2012&scsm=1&ssd=1&sort=country&ds=.&br=1&c=134&s=NGDP_R,NGDP,NGDPD,NGDPRPC,NGDPPC,NGDPDPC,PPPGDP,PPPPC&grp=0&a=). International Monetary Fund. October 2012. . Retrieved 9-October-2012.

[8] "Human development index" (http://hdr.undp.org/en/media/HDR_2011_EN_Table1.pdf). United Nations Development Programme. 2011. . Retrieved 5 November 2011.

[9] Mangold, Max, ed. (1995) (in German). *Duden, Aussprachewörterbuch* (6th ed.). Dudenverlag. pp. 271, 53f. ISBN 978-3-411-20916-3.

[10] The Latin name *Sacrum Imperium* (Holy Empire) is documented as far back as 1157. The Latin name *Sacrum Romanum Imperium* (Holy Roman Empire) was first documented in 1254. The full name "Holy Roman Empire of the German Nation" (*Heiliges Römisches Reich Deutscher Nation*) dates back to the 15th century.
Zippelius, Reinhold (2006) [1994] (in German). *Kleine deutsche Verfassungsgeschichte: vom frühen Mittelalter bis zur Gegenwart [Brief German Constitutional History: from the Early Middle Ages to the Present]* (7th ed.). Munich: Beck. p. 25. ISBN 978-3-406-47638-9.

[11] "Germany" (http://berkleycenter.georgetown.edu/resources/countries/germany). Berkley Center for Religion, Peace, and World Affairs. . Retrieved 15 December 2011.

[12] Schulze, Hagen (1998). *Germany: A New History*. Harvard University Press. p. 4. ISBN 0-674-80688-3.

[13] Lloyd, Albert L.; Lühr, Rosemarie; Springer, Otto (1998) (in German). *Etymologisches Wörterbuch des Althochdeutschen, Band II* (http://books.google.com/books?id=iKfYGNwwNVIC&pg=PA523). Göttingen: Vandenhoeck & Ruprecht. pp. 699–704. ISBN 3-525-20768-9. . (for *diutisc*) Lloyd, Albert L.; Lühr, Rosemarie; Springer, Otto (1998) (in German). *Etymologisches Wörterbuch des Althochdeutschen, Band II* (http://books.google.com/books?id=iKfYGNwwNVIC&pg=PA516). Göttingen: Vandenhoeck & Ruprecht. pp. 685–686. ISBN 3-525-20768-9. . (for *diot*)

[14] Claster, Jill N. (1982). *Medieval Experience: 300–1400*. New York University Press. p. 35. ISBN 0-8147-1381-5.

[15] Fulbrook 1991, pp. 9–13.

[16] Bowman, Alan K.; Garnsey, Peter; Cameron, Averil (2005). *The crisis of empire, A.D. 193–337*. The Cambridge Ancient History. **12**. Cambridge University Press. p. 442. ISBN 0-521-30199-8.

[17] Fulbrook 1991, p. 11.

[18] Fulbrook 1991, pp. 13–24.

[19] Nelson, Lynn Harry. *The Great Famine (1315–1317) and the Black Death (1346–1351)* (http://www.vlib.us/medieval/lectures/black_death.html). University of Kansas. . Retrieved 19 March 2011.

[20] Fulbrook 1991, p. 27.

[21] Philpott, Daniel (January 2000). "The Religious Roots of Modern International Relations". *World Politics* **52** (2): 206–245.

[22] Macfarlane, Alan (1997). *The savage wars of peace: England, Japan and the Malthusian trap*. Blackwell. p. 51. ISBN 978-0-631-18117-0.

[23] Gagliardo, G., *Reich and Nation, The Holy Roman Empire as Idea and Reality, 1763–1806*, Indiana University Press, 1980, p. 12-13.

[24] Fulbrook 1991, p. 97.

[25] Henderson, W. O. (January 1934). "The Zollverein". *History* **19** (73): 1–19. doi:10.1111/j.1468-229X.1934.tb01791.x.

[26] "Germany" (http://www.state.gov/r/pa/ei/bgn/3997.htm). U.S. Department of State. 10 November 2010. . Retrieved 26 March 2011.

[27] Black, John, ed. (2005). *100 maps*. Sterling Publishing. p. 202. ISBN 978-1-4027-2885-3.

[28] Fulbrook 1991, pp. 135, 149.

[29] Crossland, David (22 January 2008). "Last German World War I Veteran Believed to Have Died" (http://www.spiegel.de/international/germany/0,1518,530319,00.html). *Spiegel Online*. . Retrieved 25 March 2011.

[30] Lee, Stephen J. (2003). *Europe, 1890–1945*. Routledge. p. 131. ISBN 978-0-415-25455-7.

[31] Thamer, Hans-Ulrich (2003). "Beginn der nationalsozialistischen Herrschaft (Teil 2)" (http://www.bpb.de/geschichte/nationalsozialismus/dossier-nationalsozialismus/39548/gleichschaltung) (in German). *Nationalsozialismus I*. Bonn: Federal Agency for Civic Education. . Retrieved 5 April 2012. "President von Hindenburg died on 2 August 1934. The day before, the cabinet had approved a submission making Hitler his successor. The office of the president was to be dissolved and united with that of the chancellor under the name "Führer und Reichskanzler". However, this was in breach of the Enabling Act. (shortened & paraphrased)"

[32] Fulbrook 1991, pp. 156–160.

[33] Fulbrook 1991, pp. 155–158, 172–177.

[34] "Das Ermächtigungsgesetz 1933" (http://www.dhm.de/lemo/html/nazi/innenpolitik/ermaechtigungsgesetz/index.html) (in German). Deutsches Historisches Museum. . Retrieved 25 March 2011.

[35] Stackelberg, Roderick (1999). *Hitler's Germany: Origins, interpretations, legacies*. Routledge. p. 103. ISBN 978-0-415-20115-5.

[36] "Industrie und Wirtschaft" (http://www.dhm.de/lemo/html/nazi/wirtschaft/index.html) (in German). Deutsches Historisches Museum. . Retrieved 25 March 2011.

[37] Fulbrook 1991, pp. 188–189.

[38] Fulbrook 1991, pp. 190–195.

[39] Steinberg, Heinz Günter (1991) (in German). *Die Bevölkerungsentwicklung in Deutschland im Zweiten Weltkrieg: mit einem Überblick über die Entwicklung von 1945 bis 1990*. Kulturstiftung der dt. Vertriebenen. ISBN 978-3-88557-089-9.

[40] Niewyk, Donald L.; Nicosia, Francis R. (2000). *The Columbia Guide to the Holocaust*. Columbia University Press. pp. 45–52. ISBN 978-0-231-11200-0.

[41] "Leaders mourn Soviet wartime dead" (http://news.bbc.co.uk/2/hi/europe/4530565.stm). *BBC News*. 9 May 2005. . Retrieved 18 March 2011.

[42] Overy, Richard (17 February 2011). "Nuremberg: Nazis on Trial" (http://www.bbc.co.uk/history/worldwars/wwtwo/nuremberg_article_01.shtml). BBC History. . Retrieved 25 March 2011.

[43] Rűdiger Overmans. *Deutsche militärische Verluste im Zweiten Weltkrieg*. Oldenbourg 2000. ISBN 3-486-56531-1

[44] Das Deutsche Reich und der Zweite Weltkrieg, Bd. 9/1, ISBN 3-421-06236-6. Page 460 (This study was prepared by the German Armed Forces Military History Research Office, an agency of the German government)

[45] Peter Antill; Peter Dennis (10 October 2005). *Berlin 1945: end of the Thousand Year Reich* (http://books.google.com/books?id=vAzgsCDUky0C). Osprey Publishing. p. 85. ISBN 978-1-84176-915-8. . Retrieved 24 June 2011.

[46] Bonn : Kulturstiftung der Deutschen Vertriebenen, *Vertreibung und Vertreibungsverbrechen, 1945–1948 : Bericht des Bundesarchivs vom 28. Mai 1974 : Archivalien und ausgewählte Erlebnisberichte* / [Redaktion, Silke Spieler]. Bonn :1989 ISBN 3-88557-067-X. (This is a study of German expulsion casualties due to "war crimes" prepared by the German government Archives)

[47] Germany reports. With an introd. by Konrad Adenauer. Germany (West). Presse- und Informationsamt. Wiesbaden, Distribution: F. Steiner, 1961] Page 32

[48] Robert N. Proctor, *Racial Hygiene: Medicine under the Nazis*, Harvard 1988,

[49] Beevor, Antony (2003) [2002]. *Berlin: The downfall 1945*. Penguin. pp. 31–32, 409–412. ISBN 978-0-14-028696-0.

[50] Richard J. Evans, "The Other Horror, Review of *Orderly and Humane: The Expulsion of the Germans After the Second World War*, by R.M. Douglas" (http://www.tnr.com/book/review/orderly-humane-expulsion-germans-richard-evans?page=0,1&utm_source=The New Republic&utm_campaign=c58b90ef8c-TNR_BA_062812&utm_medium=email), *The New Republic*, 25 June 2012. Retrieved 1 July 2012

[51] Wise, Michael Z. (1998). *Capital dilemma: Germany's search for a new architecture of democracy*. Princeton Architectural Press. p. 23. ISBN 978-1-56898-134-5.

[52] maw/dpa (11 March 2008). "New Study Finds More Stasi Spooks" (http://www.spiegel.de/international/germany/0,1518,540771,00.html). *Spiegel Online – english site (www.spiegel.de/international)*. Der Spiegel. . Retrieved 30 October 2011. "189,000 people were informers for the Stasi – the former Communist secret police – when East Germany collapsed in 1989 – 15,000 more than previous studies had suggested. [...] about one in 20 members of the former East German Communist party, the SED, was a secret police informant."

[53] Colchester, Nico (1 January 2001). "D-mark day dawns" (http://www.ft.com/cms/s/2/504285c4-68b6-11da-bd30-0000779e2340,dwp_uuid=6f876a3c-e19f-11da-bf4c-0000779e2340.html). *Financial Times* (London). . Retrieved 19 March 2011.

[54] Protzman, Ferdinand (22 August 1989). "Westward Tide of East Germans Is a Popular No-Confidence Vote" (http://www.nytimes.com/1989/08/22/world/westward-tide-of-east-germans-is-a-popular-no-confidence-vote.html?pagewanted=all&src=pm). *www.nytimes.com* (The New York Times). . Retrieved 30 October 2011. "Behind the mass flight, Western experts say, is widespread and deepening disillusionment with the Honecker leadership's policies, particularly the refusal to consider the type of economic and political changes taking place elsewhere in the Eastern Bloc."

[55] "Gesetz zur Umsetzung des Beschlusses des Deutschen Bundestages vom 20. Juni 1991 zur Vollendung der Einheit Deutschlands" (http://bundesrecht.juris.de/berlin_bonng/index.html) (in German). Bundesministerium der Justiz. 26 April 1994. . Retrieved 19 April 2011.

[56] "Brennpunkt: Hauptstadt-Umzug" (http://www.focus.de/panorama/boulevard/brennpunkt-hauptstadt-umzug_aid_175751.html) (in German). *Focus* (Munich). 12 April 1999. . Retrieved 19 March 2011.

[57] Dempsey, Judy (31 October 2006). "Germany is planning a Bosnia withdrawal" (http://www.nytimes.com/2006/10/31/world/europe/31iht-germany.3343963.html). *International Herald Tribune* (Paris). . Retrieved 7 May 2011.

[58] Merz, Sebastian (November 2007). "Still on the way to Afghanistan? Germany and its forces in the Hindu Kush" (http://www.sipri.org/research/conflict/publications/merz) (PDF). Stockholm International Peace Research Institute. pp. 2, 3. . Retrieved 16 April 2011.
[59] "Climate in Germany" (http://www.germanculture.com.ua/library/facts/bl_climate.htm). GermanCulture. . Retrieved 26 March 2011.
[60] "Terrestrial Ecoregions" (http://wwf.panda.org/about_our_earth/ecoregions/ecoregion_list/). WWF. . Retrieved 19 March 2011.
[61] Strohm, Kathrin (May 2010). "Arable farming in Germany" (http://www.agribenchmark.org/fileadmin/freefiles/ccc_2010/Poster_Germany.pdf). Agri benchmark. . Retrieved 14 April 2011.
[62] Bekker, Henk (2005). *Adventure Guide Germany*. Hunter. p. 14. ISBN 978-1-58843-503-3.
[63] "Zoo Facts" (http://www.americanzoos.info/Zoofacts.html). Zoos and Aquariums of America. . Retrieved 16 April 2011.
[64] "Der Zoologische Garten Berlin" (http://www.zoo-berlin.de/zoo/unternehmen/historie.html) (in German). Zoo Berlin. . Retrieved 19 March 2011.
[65] "Basic Law for the Federal Republic of Germany" (https://www.btg-bestellservice.de/pdf/80201000.pdf). *Deutscher Bundestag*. Btg-bestellservice. October 2010. . Retrieved 14 April 2011.
[66] "Christian Democratic Union/Christian Social Union" (http://countrystudies.us/germany/159.htm). U.S. Library of Congress. . Retrieved 26 March 2011.
[67] "Federal Constitutional Court" (http://www.bundesverfassungsgericht.de/en/index.html). Bundesverfassungsgericht. . Retrieved 26 March 2011.
[68] "Völkerstrafgesetz Teil 1 Allgemeine Regelungen" (http://bundesrecht.juris.de/vstgb/BJNR225410002.html#BJNR225410002BJNG000200000) (in German). Bundesministerium der Justiz. . Retrieved 19 April 2011.
[69] "§ 2 Strafvollzugsgesetz" (http://www.gesetze-im-internet.de/stvollzg/__2.html) (in German). Bundesministerium der Justiz. . Retrieved 26 March 2011.
[70] Jehle, Jörg-Martin; German Federal Ministry of Justice (2009). *Criminal Justice in Germany* (http://books.google.com/books?id=-V-ng-8jOoQC&pg=PA23). Forum-Verlag. p. 23. ISBN 978-3-936999-51-8. .
[71] Casper, Gerhard; Zeisel, Hans (January 1972). "Lay Judges in the German Criminal Courts". *Journal of Legal Studies* **1** (1): 141. JSTOR 724014.
[72] The individual denomination is either *Land* [state], *Freistaat* [free state] or *Freie (und) Hansestadt* [free (and) Hanseatic city].
"The Federal States" (http://www.bundesrat.de/nn_11006/EN/organisation-en/laender-en/laender-en-node.html?__nnn=true). *www.bundesrat.de*. Bundesrat of Germany. . Retrieved 17 July 2011.
"Amtliche Bezeichnung der Bundesländer [Official denomination of federated states]" (http://www.auswaertiges-amt.de/DE/Infoservice/Terminologie/Bundeslaender/Uebersicht_node.html) (in German) (PDF; download file "Englisch"). *www.auswaertiges-amt.de*. Federal Foreign Office. . Retrieved 22 October 2011.
[73] "Example for state constitution: "Constitution of the Land of North Rhine-Westphalia"" (http://www.landtag.nrw.de/portal/WWW/GB_I/I.7/Europa/Wissenswertes/English_information/North_Rhine_Westphalia_Constitution_revised.jsp). Landtag (state assembly) of North Rhine-Westphalia. . Retrieved 17 July 2011.
[74] "Kreisfreie Städte und Landkreise nach Fläche und Bevölkerung 31.12.2010" (https://www.destatis.de/DE/ZahlenFakten/LaenderRegionen/Regionales/Gemeindeverzeichnis/Administrativ/Aktuell/04Kreise.xls;jsessionid=611904454C9287BBBD47BE5D6103CAC1.cae2?__blob=publicationFile) (in German) (XLS). Statistisches Bundesamt Deutschland. October 2011. . Retrieved 6 April 2012.
[75] "German Missions Abroad" (http://www.auswaertiges-amt.de/EN/AAmt/Auslandsvertretungen/Uebersicht_node.html). German Federal Foreign Office. . Retrieved 26 March 2011.
[76] "The Embassies" (http://www.auswaertiges-amt.de/EN/AAmt/Auslandsvertretungen/Botschaften_node.html). German Federal Foreign Office. . Retrieved 18 July 2012.
[77] "The EU budget 2011 in figures" (http://ec.europa.eu/budget/figures/2011/2011_en.cfm). European Commission. . Retrieved 6 May 2011.
[78] "United Nations regular budget for the year 2011" (http://www.un.org/ga/search/view_doc.asp?symbol=ST/ADM/SER.B/824). UN Committee on Contributions. . Retrieved 6 May 2011.
[79] "Declaration by the Franco-German Defence and Security Council" (http://www.ambafrance-uk.org/Declaration-by-the-Franco-German,4519.html). French Embassy UK. 13 May 2004. . Retrieved 19 March 2011.
[80] Freed, John C. (4 April 2008). "The leader of Europe? Answers an ocean apart" (http://www.nytimes.com/2008/04/04/world/europe/04iht-poll.4.11666423.html). *The New York Times*. . Retrieved 28 March 2011.
[81] "Aims of German development policy" (http://www.bmz.de/en/index.html). Federal Ministry for Economic Cooperation and Development. 10 April 2008. . Retrieved 26 March 2011.
[82] "Net Official Development Assistance 2009" (http://www.oecd.org/dataoecd/17/9/44981892.pdf). OECD. . Retrieved 26 March 2011.
[83] "Speech by Chancellor Angela Merkel to the United Nations General Assembly" (http://www.bundesregierung.de/nn_6566/Content/EN/Reden/2010/2010-09-21-merkel-mdg-gipfel.html). *Die Bundesregierung*. 21 September 2010. . Retrieved 18 March 2011.
[84] Harrison, Hope (2004). "American détente and German ostpolitik, 1969–1972" (http://www.ghi-dc.org/files/publications/bu_supp/supp1/supp-01_005.pdf). *Bulletin Supplement* (German Historical Institute) **1**. . Retrieved 26 March 2011.
[85] "Germany's New Face Abroad" (http://www.dw-world.de/dw/article/0,2144,1741310,00.html). *Deutsche Welle*. 14 October 2005. . Retrieved 26 March 2011.

[86] "Ready for a Bush hug?" (http://www.economist.com/node/7141311?story_id=7141311). *The Economist*. 6 July 2006. . Retrieved 19 March 2011.

[87] "U.S.-German Economic Relations Factsheet" (http://germany.usembassy.gov/germany/img/assets/9336/econ_factsheet_may2006.pdf). U.S. Embassy in Berlin. May 2006. . Retrieved 26 March 2011.

[88] "The 15 countries with the highest military expenditure in 2011" (http://www.sipri.org/research/armaments/milex/resultoutput/milex_15/the-15-countries-with-the-highest-military-expenditure-in-2011-table/view). Stockholm International Peace Research Institute. September 2011. . Retrieved 7 April 2012.

[89] "Grundgesetz für die Bundesrepublik Deutschland, Artikel 65a,87,115b" (http://www.gesetze-im-internet.de/bundesrecht/gg/gesamt.pdf) (in German). Bundesministerium der Justiz. . Retrieved 19 March 2011.

[90] "Die Stärke der Streitkräfte" (http://www.bundeswehr.de/portal/a/bwde/!ut/p/c4/DcmxDYAwDATAWVgg7unYAugc8kSWI4OMIesTXXm002D8SeWQy7jRStshc-4p94L0hENCnXEGUvXXSuMKG8FwBd26TD9uIZiT/) (in German). Bundeswehr. 23 March 2012. . Retrieved 20 April 2012.

[91] "Ausblick: Die Bundeswehr der Zukunft" (http://www.bundeswehr.de/portal/a/bwde/!ut/p/c4/04_SB8K8xLLM9MSSzPy8xBz9CP3I5EyrpHK9pPKUVL3ikqLUzJLsosTUtJJUvbzU0vTU4pLEnJLSvHRUuYKcxDygoH5BtqMiAMTJdF8!/) (in German). Bundeswehr. . Retrieved 5 June 2011.

[92] "Einsatzzahlen – Die Stärke der deutschen Einsatzkontingente" (http://www.bundeswehr.de/portal/a/bwde/einsaetze/einsatzzahlen?yw_contentURL=/C1256EF4002AED30/W264VFT2439INFODE/content.jsp) (in German). Bundeswehr. . Retrieved 14 April 2011.

[93] Connolly, Kate (22 November 2010). "Germany to abolish compulsory military service" (http://www.guardian.co.uk/world/2010/nov/22/germany-abolish-compulsory-military-service). *The Guardian* (UK). . Retrieved 7 April 2011.

[94] Pidd, Helen (16 March 2011). "Marching orders for conscription in Germany, but what will take its place?" (http://www.guardian.co.uk/world/2011/mar/16/conscription-germany-army). *The Guardian* (UK). . Retrieved 7 April 2011.

[95] "Frauen in der Bundeswehr" (http://www.bundeswehr.de/portal/a/bwde/!ut/p/c4/FcwxEoUgDAXAE0l6O0_x1YZ5QMSMEp2In-urs_3STC_FXzKqHIqdRpqi9KG50BK7qxpL3Qy8VHbZbk07MqtbDDerF_WJzYdGv286DbmAJj26iLgynaUM) (in German). Bundeswehr. . Retrieved 14 April 2011.

[96] Norris, Floyd (20 February 2010). "A Shift in the Export Powerhouses" (http://www.nytimes.com/2010/02/20/business/economy/20charts.html). *The New York Times*. . Retrieved 27 March 2011.

[97] "CPI 2009 table" (http://archive.transparency.org/policy_research/surveys_indices/cpi/2009/cpi_2009_table). Transparency International. . Retrieved 15 May 2012.

[98] "The Innovation Imperative in Manufacturing: How the United States Can Restore Its Edge" (http://www.bcg.com/documents/file15445.pdf). Boston Consulting Group. March 2009. . Retrieved 19 March 2011.

[99] "Gross domestic product (2009)" (http://siteresources.worldbank.org/DATASTATISTICS/Resources/GDP.pdf). *The World Bank: World Development Indicators database*. World Bank. 27 September 2010. . Retrieved 1 January 2011.
Field listing – GDP (official exchange rate) (https://www.cia.gov/library/publications/the-world-factbook/fields/2195.html)

[100] "Gross domestic product (2009)" (http://siteresources.worldbank.org/DATASTATISTICS/Resources/GDP_PPP.pdf). *The World Bank: World Development Indicators database*. World Bank. 27 September 2010. . Retrieved 5 October 2010.
Field listing – GDP (PPP exchange rate) (https://www.cia.gov/library/publications/the-world-factbook/rankorder/2001rank.html)

[101] Financial Crisis. "EU budget: who pays what and how it is spent" (http://www.telegraph.co.uk/finance/financialcrisis/9643193/EU-budget-who-pays-what-and-how-it-is-spent.html). Telegraph. . Retrieved 2012-11-04.

[102] "The Labour Market in May 2012: Positive Underlying Trend Weakens" (http://www.arbeitsagentur.de/nn_426140/EN/zentraler-Content/Presse/Presse-12-023-EN-690148.html). .

[103] Press office of the Deutsche Bundesbank. "Deutsche Bundesbank - Statistics" (http://www.bundesbank.de/statistik/statistik_konjunktur.en.php). Bundesbank.de. . Retrieved 4 June 2012.

[104] Andrews, Edmund L. (1 January 2002). "Germans Say Goodbye to the Mark, a Symbol of Strength and Unity" (http://www.nytimes.com/2002/01/01/world/germans-say-goodbye-to-the-mark-a-symbol-of-strength-and-unity.html). *The New York Times*. . Retrieved 18 March 2011.

[105] Taylor Martin, Susan (28 December 1998). "On Jan. 1, out of many arises one Euro". *St. Petersburg Times*: p. National, 1.A.

[106] Berg, S.; Winter, S.; Wassermann, A. (5 September 2005). "The Price of a Failed Reunification" (http://www.spiegel.de/international/spiegel/0,1518,373639,00.html). *Spiegel Online*. . Retrieved 28 November 2006.

[107] Kulish, Nicholas (19 June 2009). "In East Germany, a Decline as Stark as a Wall" (http://www.nytimes.com/2009/06/19/world/europe/19germany.html). *The New York Times*. . Retrieved 27 March 2011.

[108] "Germany agrees on 50-billion-euro stimulus plan" (http://www.france24.com/en/20090106-germany-agrees-new-50-billion-euro-stimulus-plan). *France 24*. 6 January 2009. . Retrieved 27 March 2011.

[109] "The 100 Top Brands 2010" (http://www.interbrand.com/en/best-global-brands/best-global-brands-2008/best-global-brands-2010.aspx). Interbrand. . Retrieved 27 March 2011.

[110] Gavin, Mike (23 September 2010). "Germany Has 1,000 Market-Leading Companies, Manager-Magazin Says" (http://www.businessweek.com/news/2010-09-23/germany-has-1-000-market-leading-companies-manager-magazin-says.html). *Businessweek* (New York). . Retrieved 27 March 2011.

[111] "Global 500: Countries – Germany" (http://money.cnn.com/magazines/fortune/global500/2010/countries/Germany.html). *Forbes*. 26 July 2010. . Retrieved 27 March 2011.

[112] "Autobahn-Temporegelung" (http://www.presse.adac.de/standpunkte/Verkehr/Autobahn_Temporegelung.asp?active1=tcm:11-18784-4) (in German) (Press release). ADAC. June 2010. . Retrieved 19 March 2011.

[113] "Geschäftsbericht 2006" (http://web.archive.org/web/20070809140315/http://www.db.de/site/bahn/de/unternehmen/investor__relations/finanzberichte/geschaeftsbericht/geschaeftsbericht__2006.html) (in German). Deutsche Bahn. Archived from the original (http://www.db.de/site/bahn/de/unternehmen/investor__relations/finanzberichte/geschaeftsbericht/geschaeftsbericht__2006.html) on 9 August 2007. . Retrieved 27 March 2011.

[114] "Airports in Germany" (http://www.aircraft-charter-world.com/airports/europe/germany.htm). Air Broker Center International. . Retrieved 16 April 2011.

[115] "BER will open on 17 March 2013" (http://preview.berlin-airport.de/en/company/about-us/news/2012/2012-05-17-BER-will-open-on-17-march-2013/index.php). Berlin Brandenburg Airport. . Retrieved 31 May 2012.

[116] "Overview/Data: Germany" (http://www.eia.gov/countries/country-data.cfm?fips=GM). U.S. Energy Information Administration. 30 June 2010. . Retrieved 19 April 2011.

[117] "Energy imports, net (% of energy use)" (http://data.worldbank.org/indicator/EG.IMP.CONS.ZS). The World Bank Group. . Retrieved 18 April 2011.

[118] Reuters Berlin (7 July 2010). "* Environment * Renewable energy Germany targets switch to 100% renewables for its electricity by 2050" (http://www.guardian.co.uk/environment/2010/jul/07/germany-renewable-energy-electricity). *The Guardian* (UK). . Retrieved 18 April 2011.

[119] "Primärenergieverbrauch nach Energieträgern" (http://www.bmwi.de/BMWi/Redaktion/Binaer/Energiedaten/energiegewinnung-und-energieverbrauch2-primaerenergieverbrauch.xls) (in German). Bundesministerium für Wirtschaft und Technologie. December 2010. . Retrieved 18 April 2011.

[120] "Germany split over green energy" (http://news.bbc.co.uk/2/hi/europe/4295389.stm). *BBC News*. 25 February 2005. . Retrieved 27 March 2011.

[121] "Deutschland erfüllte 2008 seine Klimaschutzverpflichtung nach dem Kyoto-Protokoll" (http://www.kst.portalu.de/dokumente/Aktuelles/Inventar-Bericht_Treibhausgas_2008.pdf) (in German) (Press release). Umweltbundesamt. 1 February 2010. . Retrieved 19 March 2011.

[122] "Germany greenest country in the world" (http://articles.timesofindia.indiatimes.com/2008-06-21/pollution/27752791_1_energy-security-global-energy-germany). *The Times of India* (New Delhi). 21 June 2008. . Retrieved 26 March 2011.

[123] "Record High 2010 Global Carbon Dioxide Emissions from Fossil-Fuel Combustion and Cement Manufacture Posted on CDIAC Site" (http://cdiac.ornl.gov/trends/emis/prelim_2009_2010_estimates.html). Carbon Dioxide Information Analysis Center. . Retrieved 15 May 2012.

[124] "Federal Report on Research and Innovation 2010" (http://www.bmbf.de/pub/bufi_2010_en.pdf) (PDF). Federal Ministry of Education and Research. June 2010. . Retrieved 15 May 2012.

[125] "Nobel Prize" (http://nobelprize.org/). Nobelprize.org. . Retrieved 27 March 2011.

[126] "Swedish academy awards" (http://www.sciencenews.org/view/generic/id/63944/title/Swedish_academy_awards). *ScienceNews*. . Retrieved 1 October 2010.

[127] National Science Nobel Prize shares 1901–2009 by citizenship at the time of the award (http://www.idsia.ch/~juergen/sci.html) and by country of birth (http://www.idsia.ch/~juergen/scinat.html). From Schmidhuber, J. (2010). "Evolution of National Nobel Prize Shares in the 20th century" (http://www.idsia.ch/~juergen/nobelshare.html). . Retrieved 27 March 2011.

[128] Roberts, J. M. (2002). *The New Penguin History of the World*. Allen Lane. p. 1014. ISBN 978-0-7139-9611-1.

[129] "The First Nobel Prize" (http://www.dw-world.de/dw/article/0,,5984670,00.html). *Deutsche Welle*. 8 September 2010. . Retrieved 27 March 2011.

[130] "Otto Hahn" (http://www.famousscientists.org/otto-hahn). FamousScientists.org. . Retrieved 15 December 2011.

[131] "Gottfried Wilhelm Leibniz Prize" (http://web.archive.org/web/20080621091621/http://www.dfg.de/en/research_funding/scientific_prizes/gw_leibniz_prize.html). DFG. Archived from the original (http://www.dfg.de/en/research_funding/scientific_prizes/gw_leibniz_prize.html) on 21 June 2008. . Retrieved 27 March 2011.

[132] Bianchi, Luigi. "The Great Electromechanical Computers" (http://www.yorku.ca/lbianchi/sts3700b/lecture17a.html). York University. . Retrieved 17 April 2011.

[133] "The Zeppelin" (http://www.centennialofflight.gov/essay/Lighter_than_air/zeppelin/LTA8.htm). U.S. Centennial of Flight Commission. . Retrieved 27 March 2011.

[134] "Historical figures in telecommunications" (http://www.itu.int/aboutitu/HistoricalFigures.html). International Telecommunication Union. 14 January 2004. . Retrieved 27 March 2011.

[135] Roland Berger Strategy Consultants: *Green Growth, Green Profit – How Green Transformation Boosts Business* Palgrave Macmillan, New York 2010, ISBN 978-0-230-28543-9

[136] "Country Comparison :: Population" (https://www.cia.gov/library/publications/the-world-factbook/rankorder/2119rank.html?countryName=Lebanon&countryCode=le&regionCode=me&rank=126). CIA. . Retrieved 26 June 2011.

[137] "Demographic Transition Model" (http://geographyfieldwork.com/DemographicTransition.htm). Barcelona Field Studies Centre. 27 September 2009. . Retrieved 28 March 2011.

[138] "Im Jahr 2060 wird jeder Siebente 80 Jahre oder älter sein" (https://www.destatis.de/DE/PresseService/Presse/Pressemitteilungen/2009/11/PD09_435_12411.html;jsessionid=7247D871DEB96E4AE313EE55A45CEFC1.cae1) (in German) (Press release). Destatis. 18 November 2009. . Retrieved 6 April 2012.
Details on the methodology, detailed tables, etc. are provided at "Bevölkerungsentwicklung in Deutschland bis 2060" (http://web.archive.org/web/20101229190016/http://www.destatis.de/jetspeed/portal/cms/Sites/destatis/Internet/DE/Presse/abisz/Bevoelkerungsvorausberechnung,templateId=renderPrint.psml) (in German). Statistisches Bundesamt. 18 November 2009. Archived from the original (http://www.destatis.de/jetspeed/portal/cms/Sites/destatis/Internet/DE/Presse/abisz/Bevoelkerungsvorausberechnung,templateId=renderPrint.psml) on 29 December 2010. . Retrieved 15 February 2011.

[139] Population and employment: Population with migrant background – Results of the 2010 microcensus (https://www.destatis.de/DE/Publikationen/Thematisch/Bevoelkerung/MigrationIntegration/Migrationshintergrund2010220107004.pdf) The Federal Statistical Office defines persons with a migrant background as all persons who migrated to the present area of the Federal Republic of Germany after 1949, plus all foreign nationals born in Germany and all persons born in Germany as German nationals with at least one parent who migrated to Germany or was born in Germany as a foreign national.

[140] "Publikation - STATmagazin - Population - Families with a migrant background: traditional values count - Federal Statistical Office (Destatis)" (https://www.destatis.de/EN/Publications/STATmagazin/Population/2012_03/2012_03MigrantBackground.html). Destatis.de. 2012-03-13. . Retrieved 2012-11-04.

[141] "International Migration 2006" (http://www.un.org/esa/population/publications/2006Migration_Chart/Migration2006.pdf). UN Department of Economic and Social Affairs. . Retrieved 18 March 2011.

[142] "Germany" (http://www.focus-migration.de/Germany.1509.0.html?&L=1). Focus-Migration. . Retrieved 28 March 2011.

[143] "20% of Germans have immigrant roots". *Burlington Free Press*: p. 4A. 15 July 2010.

[144] "Bevölkerung nach Migrationshintergrund" (http://web.archive.org/web/20101229095012/http://www.destatis.de/jetspeed/portal/cms/Sites/destatis/Internet/DE/Content/Statistiken/Bevoelkerung/MigrationIntegration/Migrationshintergrund/Tabellen/Content100/MigrationshintergrundStaatsangehoerigkeit,templateId=renderPrint.psml) (in German). German Federal Statistical Office. Archived from the original (http://www.destatis.de/jetspeed/portal/cms/Sites/destatis/Internet/DE/Content/Statistiken/Bevoelkerung/MigrationIntegration/Migrationshintergrund/Tabellen/Content100/MigrationshintergrundStaatsangehoerigkeit,templateId=renderPrint.psml) on 29 December 2010. . Retrieved 28 March 2011.

[145] "Fewer Ethnic Germans Immigrating to Ancestral Homeland" (http://www.migrationinformation.org/Feature/display.cfm?ID=201). Migration Information Source. February 2004. . Retrieved 28 March 2011.

[146] "Bevölkerung mit Migrationshintergrund I | bpb" (http://www.bpb.de/nachschlagen/zahlen-und-fakten/soziale-situation-in-deutschland/61646/migrationshintergrund-i) (in (German)). Bpb.de. 6 October 2011. . Retrieved 13 August 2012.

[147] "Pressemitteilungen - Statistisches Bundesamt (Destatis) - Ein Fünftel der Bevölkerung in Deutschland hatte 2010 einen Migrationshintergrund- Statistisches Bundesamt (Destatis)" (https://www.destatis.de/DE/PresseService/Presse/Pressemitteilungen/2011/09/PD11_355_122.html) (in (German)). Destatis.de. 26 September 2011. . Retrieved 13 August 2012.

[148] "Regionales Monitoring 2010 - Daten und Karten zu den Europäischen Metropolregionen in Deutschland" (http://www.deutsche-metropolregionen.org/fileadmin/ikm/01_monitoring/IKM_Monitoring2010.pdf) (in German). Bundesamt für Bauwesen und Raumordnung. 2010. p. 10. . Retrieved 11 April 2012.

[149] "EKD-Statistik: Christen in Deutschland 2007" (http://www.ekd.de/statistik/mitglieder.html) (in German). Evangelische Kirche in Deutschland. . Retrieved 28 March 2011.

[150] "Konfessionen in Deutschland" (http://fowid.de/fileadmin/datenarchiv/Religionszugehoerigkeit_Bevoelkerung__1950-2008.pdf) (in German). Fowid. 9 September 2009. . Retrieved 28 March 2011.

[151] "Chapter 2: Wie viele Muslime leben in Deutschland?" (http://www.bmi.bund.de/cae/servlet/contentblob/566008/publicationFile/31710/vollversion_studie_muslim_leben_deutschland_.pdf;jsessionid=6B8CD26E2AC179111AF4F75650B84B1A) (in German) (PDF). *Muslimisches Leben in Deutschland*. Bundesamt für Migration und Flüchtlinge. June 2009. pp. 80, 97. ISBN 978-3-9812115-1-1. . Retrieved 28 March 2011.

[152] "Religionen in Deutschland: Mitgliederzahlen" (http://www.remid.de/remid_info_zahlen.htm) (in German). Religionswissenschaftlicher Medien- und Informationsdienst. 31 October 2009. . Retrieved 28 March 2011.

[153] Blake, Mariah (10 November 2006). "In Nazi cradle, Germany marks Jewish renaissance" (http://www.csmonitor.com/2006/1110/p25s02-woeu.html). *Christian Science Monitor*. . Retrieved 28 March 2011.

[154] Schnabel, U. (15 March 2007). "Buddhismus Eine Religion ohne Gott" (http://www.zeit.de/2007/12/Buddhismus) (in German). *Die Zeit* (Hamburg). . Retrieved 19 March 2011.

[155] European Commission (2006). "Special Eurobarometer 243: Europeans and their Languages (Survey)" (http://ec.europa.eu/public_opinion/archives/ebs/ebs_243_en.pdf). Europa (web portal). . Retrieved 28 March 2011.
European Commission (2006). "Special Eurobarometer 243: Europeans and their Languages (Executive Summary)" (http://ec.europa.eu/public_opinion/archives/ebs/ebs_243_sum_en.pdf). Europa (web portal). . Retrieved 28 March 2011.

[156] European Commission (2004). "Many tongues, one family. Languages in the European Union" (http://ec.europa.eu/publications/booklets/move/45/en.pdf). Europa (web portal). . Retrieved 28 March 2011.

[157] "Sprechen Sie Deutsch?" (http://www.economist.com/node/15731354). *The Economist*. 18 March 2010. . Retrieved 16 April 2011.

[158] Grotlüschen, Anke; Riekmann, Wibke (2011). "leo.- Level One Survey. Presseheft" (http://www.vhs-sachsen.de/uploads/media/Executive_Summary_leo_v8_Journalismusfassung.pdf) (in German). University of Hamburg. . Retrieved 15 April 2011.

[159] "Country profile: Germany" (http://lcweb2.loc.gov/frd/cs/profiles/Germany.pdf). Library of Congress. April 2008. . Retrieved 28 March 2011.

[160] "The Educational System in Germany" (http://academic.cuesta.edu/intlang/german/education.html). Cuesta College. 31 August 2002. . Retrieved 16 May 2011.

[161] "Top 100 World Universities" (http://web.archive.org/web/20080822124509/http://www.arwu.org/rank2008/ARWU2008_A(EN).htm). Academic Ranking of World Universities. Archived from the original (http://www.arwu.org/rank2008/ARWU2008_A(EN).htm) on 22 August 2008. . Retrieved 28 March 2011.

[162] "Tuition Fees at university in Germany" (http://www.studyineurope.eu/study-in-germany/admission/tuition-fees). StudyinEurope.eu. 2009. . Retrieved 19 March 2011.

[163] *Health Care Systems in Transition: Germany* (http://www.euro.who.int/__data/assets/pdf_file/0010/80776/E68952.pdf). European Observatory on Health Care Systems. 2000. p. 8. AMS 5012667 (DEU). . Retrieved 15 April 2011.

[164] "Die Gesundheitsreform 2007: Was hat sich geändert?" (http://www.krankenkassen.de/gesetzliche-krankenkassen/gesundheitsreform/Gesundheitsreform-zentrale-Punkte-Fragen/). Krankenkassen.de. . Retrieved 16 April 2012.

[165] "Core Health Indicators" (http://apps.who.int/whosis/database/core/core_select.cfm). World Health Organization. . Retrieved 6 June 2011.

[166] "2010: Herz-/Kreislauferkrankungen verursachen 41 % aller Todesfälle" (https://www.destatis.de/DE/ZahlenFakten/GesellschaftStaat/Gesundheit/Todesursachen/Aktuell.html;jsessionid=B6BD78C7EDB6FE6693DE58EA7B23D148.cae1) (in German). Destatis.de. . Retrieved 6 April 2012.

[167] "Country Profile Germany" (http://lcweb2.loc.gov/frd/cs/profiles/Germany.pdf) (PDF). Library of Congress Federal Research Division. April 2008. . Retrieved 7 May 2011.
This article may incorporate text from this source, which is in the public domain.

[168] Wasser, Jeremy (6 April 2006). "Spätzle Westerns" (http://www.spiegel.de/international/0,1518,410135,00.html). *Spiegel Online International*. . Retrieved 28 March 2011.

[169] "Unbelievable Multitude" (http://www.dw-world.de/dw/0,,8009,00.html). *Deutsche Welle*. . Retrieved 28 March 2011.

[170] "World Heritage Sites in Germany" (http://www.worldheritagesite.org/countries/germany.html). UNESCO. . Retrieved 3 October 2010.

[171] "Human Development Report 2010 Table 4 Gender Inequality Index" (http://hdr.undp.org/en/media/HDR_2010_EN_Tables_reprint.pdf). United Nations Development Programme. pp. 156–160. . Retrieved 20 April 2011.

[172] "Germany extends gay rights" (http://www.news24.com/World/News/Germany-extends-gay-rights-20041029). News24. 29 October 2004. . Retrieved 19 March 2011.

[173] Heckmann, Friedrich (2003). *The Integration of Immigrants in European Societies: national differences and trends of convergence*. Lucius & Lucius. pp. 51 ff. ISBN 978-3-8282-0181-1.

[174] "2010 Anholt-GfK Roper Nation Brands Index" (http://www.gfk.com/group/press_information/press_releases/006688/index.en.html) (Press release). GfK. 12 October 2010. . Retrieved 28 March 2011.

[175] "Views of US Continue to Improve in 2011 BBC Country Rating Poll" (http://www.worldpublicopinion.org/pipa/articles/views_on_countriesregions_bt/680.php?nid=&id=&pnt=680&lb=#ger). *Worldpublicopinion.org*. 7 March 2011. . Retrieved 28 March 2011.

[176] *A Dictionary of Architecture and Landscape Architecture*. Oxford University Press. 2006. p. 880. ISBN 0-19-860678-8.

[177] (13 April 2011). "Bundesverband Musikindustrie: Deutschland drittgrößter Musikmarkt weltweit" (http://www.musikindustrie.de/politik_einzelansicht0/back/110/news/jahrespressekonferenz-des-bvmi/). Musikindustrie.de. . Retrieved 28 February 2012.

[178] Espmark, Kjell (3 December 1999). "The Nobel Prize in Literature" (http://nobelprize.org/nobel_prizes/literature/articles/espmark/index.html). Nobelprize.org. . Retrieved 28 March 2011.

[179] "Land of ideas" (http://land-der-ideen.matrix.de/CDA/facts_printing,4563,0,,en.html). Land-der-ideen.matrix.de. . Retrieved 19 March 2011.

[180] Weidhaas, Peter; Gossage, Carolyn; Wright, Wendy A. (2007). *A History of the Frankfurt Book Fair*. Dundurn Press Ltd.. pp. 11 ff. ISBN 978-1-55002-744-0.

[181] Searle, John (1987). "Introduction". *The Blackwell Companion to Philosophy*. Wiley-Blackwell.

[182] Bordwell, David; Thompson, Kristin (2003) [1994]. "The Introduction of Sound". *Film History: An Introduction* (2nd ed.). McGraw-Hill. p. 204. ISBN 978-0-07-115141-2.

[183] "Rainer Werner Fassbinder" (http://www.fassbinderfoundation.de/node.php/en/home). Fassbinder Foundation. . Retrieved 28 March 2011.

[184] "2006 FIAPF accredited Festivals Directory" (http://www.fiapf.org/pdf/2006accreditedFestivalsDirectory.pdf). International Federation of Film Producers Associations. . Retrieved 28 March 2011.

[185] "Awards:Das Leben der Anderen" (http://www.imdb.com/title/tt0405094/awards). IMDb. . Retrieved 28 March 2011.

[186] "Country profile: Germany" (http://news.bbc.co.uk/2/hi/europe/country_profiles/1047864.stm). *BBC News*. . Retrieved 28 March 2011.

[187] "Guide to German Hams and Sausages" (http://www.germanfoods.org/consumer/facts/guidetoham.cfm). German Foods North America. . Retrieved 17 May 2012.

[188] "Germany Country Profiles for Organic Agriculture" (http://www.fao.org/organicag/display/work/display_2.asp?country=DEU&lang=en&disp=summaries). Food and Agriculture Organization. . Retrieved 6 May 2011.

[189] Schneibel, Gerhard (23 April 2010). "Brewers not worried by beer consumption drop" (http://www.dw.de/dw/article/0,,5489225,00.html). *Deutsche Welle* (Bonn). . Retrieved 6 April 2012.

[190] "Schnitzel Outcooks Spaghetti in Michelin Guide" (http://www.dw.de/dw/article/0,,2914502_page_0,00.html). *Deutsche Welle* (Bonn). 15 November 2007. . Retrieved 6 April 2012.

[191] "German cuisine beats Italy, Spain in gourmet stars" (http://in.reuters.com/article/2007/11/14/us-germany-food-idINL1447732320071114). *Reuters*. 28 March 2011. . Retrieved 19 March 2011.

[192] "Germany Info: Culture & Life: Sports" (http://www.germany.info/relaunch/culture/life/sports.html). Germany Embassy in Washington, D.C. . Retrieved 28 March 2011.

[193] Ornstein, David (23 October 2006). "What we will miss about Michael Schumacher" (http://www.guardian.co.uk/sport/2006/oct/23/formulaone.sport). *The Guardian* (UK). . Retrieved 19 March 2011.

[194] "Beijing 2008 Medal Table" (http://www.olympic.org/medallists-results?athletename=&category=343488&games=1333952&sport=&event=&mengender=false&womengender=false&mixedgender=false&teamclassification=false&individualclassification=false&continent=&country=&goldmedal=true&silvermedal=true&bronzemedal=true&worldrecord=false&olympicrecord=false&targetresults=true). International Olympic Committee. . Retrieved 19 March 2011.

[195] "Turin 2006 Medal Table" (http://www.olympic.org/medallists-results?athletename=&category=343486&games=1334152&sport=&event=&mengender=false&womengender=false&mixedgender=false&teamclassification=false&individualclassification=false&continent=&country=&goldmedal=true&silvermedal=true&bronzemedal=true&worldrecord=false&olympicrecord=false&targetresults=true&sortorder=medal&sortorder=country). International Olympic Committee. . Retrieved 19 March 2011.

Work cited

Fulbrook, Mary (1991). *A Concise History of Germany*. Cambridge University Press. ISBN 978-0-521-36836-0.

External links

- Germany profile (http://www.bbc.co.uk/news/world-europe-17299607) from the BBC News
- deutschland.de (http://www.deutschland.de/en/home-page.html?tx_fdfxyaml_pi1) – Official Germany portal (non-profit)
- Wikimedia Atlas of Germany
- Deutsche Welle (http://www.dw-world.de/) – Germany's international broadcaster
- Germany (https://www.cia.gov/library/publications/the-world-factbook/geos/gm.html) entry at *The World Factbook*
- Germany (http://ucblibraries.colorado.edu/govpubs/for/germany.htm) at *UCB Libraries GovPubs*
- Germany (http://www.dmoz.org/Regional/Europe/Germany/) at the Open Directory Project
- Facts about Germany (http://www.tatsachen-ueber-deutschland.de/en/) – by the German Federal Foreign Office
- Destatis.de (https://www.destatis.de/EN/Homepage.html) – Federal Statistical Office Germany
- Geographic data related to Germany (http://www.openstreetmap.org/browse/relation/51477) at OpenStreetMap
- Key Development Forecasts for Germany (http://www.ifs.du.edu/ifs/frm_CountryProfile.aspx?Country=DE) from International Futures

Government

- Official site of the Government (http://www.bundesregierung.de/Webs/Breg/EN/Homepage/_node.html)
- Official site of the Federal Presidents (http://www.bundespraesident.de/EN/Home/home_node.html)
- Official site of German Chancellor (http://www.bundeskanzlerin.de/Webs/BK/En/Homepage/home.html)

Tourism

- Official Germany Tourism Website (http://www.germany.travel/en/index.html)

Article Sources and Contributors

Waltersdorf, Thuringia *Source*: http://en.wikipedia.org/w/index.php?title=Waltersdorf%2C_Thuringia *Contributors*: Ksnow, Markussep

Gemeinde *Source*: http://en.wikipedia.org/w/index.php?title=Gemeinde *Contributors*: Borgx, Chanheigeorge, CharlotteWebb, DerBorg, Dirknachbar, F.chiodo, Grafen, IZAK, Jurema Oliveira, Mai-Sachme, Markussep, Moonraker12, Nyttend, PaulGarner, R9tgokunks, Radagast83, Sp4rr0W, TheMoth&TheButterfly, Tobias Conradi, TurkChan, Vuurcifer, Wikid77, Woohookitty, Xtv, Yakoo, 10 anonymous edits

Municipalities of Germany *Source*: http://en.wikipedia.org/w/index.php?title=Municipalities_of_Germany *Contributors*: Acntx, Agathoclea, Carlossuarez46, Chanheigeorge, David Kernow, DerBorg, Ehrlich91, EoGuy, Florian Blaschke, Kdammers, Mschiffler, Poulpy, Reinyday, Roke, Sardanaphalus, Stemonitis, Sundar1, Tobias Conradi, Valentinian, Worobiew, Yngvadottir, 3 anonymous edits

Weilheim an der Teck *Source*: http://en.wikipedia.org/w/index.php?title=Weilheim_an_der_Teck *Contributors*: Angr, Ksnow, Markussep, Ser Amantio di Nicolao, 2 anonymous edits

Regierungsbezirk *Source*: http://en.wikipedia.org/w/index.php?title=Regierungsbezirk *Contributors*: (, Agathoclea, Ahoerstemeier, Andrew c, Angr, Badanedwa, Bermicourt, CalJW, Capitonym, Chanheigeorge, Crusoe8181, David Parker, Delirium, DerBorg, Djmutex, Docu, Donald Albury, Dvavasour, Ehrlich91, Elapsed, Eugene van der Pijll, Exec, Frokor, Funandtrvl, Gereonmc, Green Giant, GregorB, Ingolfson, JPosten, JeLuF, KaiKemmann, Kelisi, Kwamikagami, Lfh, LilHelpa, Ludger1961, MacTire02, Marek69, MegA, Meursault2004, Mrmuk, NeunZehnHundertFünf, Numbo3, Olessi, Picapica, Piotrus, Ratzer, Reywas92, Rje, Robofish, Rossami, Royboycrashfan, Saintswithin, Sandman, Sangjinhwa, Timwi, Tom, Ulf Heinsohn, Unyoyega, Worobiew, Δ, 30 anonymous edits

Baden-Württemberg *Source*: http://en.wikipedia.org/w/index.php?title=Baden-W%C3%BCrttemberg *Contributors*: -Ilhador-, 52 Pickup, Addshore, Aec is away, Ahoerstemeier, Akso, Alan Liefting, Amberroom, Anathemius, AndrewRT, Angr, Arakunem, Aufbauten, Aviad2001, AxG, Axolotl Nr.733, Baldhur, Barticus88, Ben-Zin, BernhardtP, Bigbossfarin, Bomac, Brion VIBBER, CalicoCatLover, Calliopejen1, Caltas, Can't sleep, clown will eat me, Carabinieri, CharlotteWebb, Chl, CieloEstrellado, ClickRick, Colonies Chris, ColoradoZ, Comperr, Conversion script, Cs32en, Css, Curps, D6, DARTH SIDIOUS 2, DE, David Liuzzo, David Parker, Dawn Bard, Dbenbenn, Der Bruzzla, Der Eberswalder, DerHexer, Derflipper, Dewritech, Difluoroethene, Djmutex, Doc Quintana, Doco, Docu, Dodo19, Dofuv, Dominictimms, Domino theory, Dr. Blofeld, Edward, Ein-Rat-7000, Emrrans, Epbr123, Erianna, Fabartus, Firstorm, Flammingo, Freiwilliger, Frietjes, Funnyhat, Gadget850, Gamma, Gargolla, Gianfranco, Goldfishbutt, Good Olfactory, Graham87, Greenshed, Grenavitar, Ground Zero, H005, Haham hanuka, Hairy Dude, Hamiltondaniel, Hanberke, Hans555, Hayden120, Hebrides, Heimstern, Hgav, IZAK, Iamiyouareyou, Imars, Imladros, Iridescent, Isarra (HG), IslandGyrl, J-roc6669, Jack Bufalo Head, Jarry1250, JeLuF, Jeronimo, Jiang, Jimmy Slade, Jmalicki, John K, Joseph Solis in Australia, JuergenL, Kato2k6, Kbdank71, Kelisi, Kelownian, KevinOKeeffe, KingSlash, Kingjeff, Kintetsubuffalo, KirkEN, Koavf, Komischn, Kompakt, Konstantin, Kucharek, Kummi, Kusma, LA2, LMB, Lareusop, Laureusop, Lazar Taxon, Lazywoodz, Lear 21, Legendsword, Leszek Jańczuk, Lfh, LiterallySimon, LittleT889, Lloookkkcccnnn, Lockesdonkey, Lucky strike10, Lupo, M-le-mot-dit, Madchen Hoch, Markjoseph125, Markussep, Matthead, Melchoir, Merovingian, Meursault2004, Mezaco, MichaelJanich, Misiekuk, Mmounties, Moilleadóir, Morwen, Mr clevver, Musicandnintendo, NawlinWiki, Neckintosh, Net-net, Nico, Nlnnet, NoPuzzleStranger, Numbo3, Ohconfucius, Oknazevad, Olessi, Olivier, Omnipaedista, Orange Suede Sofa, Ospalh, P. S. F. Freitas, Paine Ellsworth, PanchoS, Paul-L, Phil Bastian, Pit, Poeloq, Pol098, QuartierLatin1968, R'n'B, Raamin, RazielZero, Readro, Reconsider the static, Renessaince, Rich Farmbrough, Richie, Rickard Vogelberg, Rif Winfield, Rillian, Rjwilmsi, RoDeWo, Rodemont2, Rosenzweig, Rosh2610, Ruhrjung, Ryn78, Sablier, Sakhani, Sandman, Saxbryn, Schneelocke, SchreiberBike, Schwab7000, Shahsan, Shredog, Skier Dude, Skierpage, Snowdog, Squash Racket, Stewartadcock, Stovl, Suite96, Sunflower129, Targaryen, Template namespace initialisation script, Themfromspace, Thingg, Tide rolls, Tillwe, TimVickers, Timwi, Tresckow, Trialsanderrors, Trickstar, Tschild, Ulf Heinsohn, Ulmensis, Unyoyega, Urban XII, Uujeroi, Valentinian, Vdiest, Vdjj1960, Waelder, Walden69, Walter Görlitz, Westee, WhyNotFreedom, Wimt, Writtenright, Wzwz, Xanderliptak, Xiiizej, Yoda1893, Yoghurt, 284 anonymous edits

List of twin towns and sister cities in Germany *Source*: http://en.wikipedia.org/w/index.php?title=List_of_twin_towns_and_sister_cities_in_Germany *Contributors*: 2000, 52 Pickup, Abdull, Acjelen, Affe, Aivazovsky, Ajbenj, Amwyll Rwden, AndTheCrowdGoesWild, Assaf050, Atabəy, Attilios, AxelBoldt, Azero III, Bender235, Berekniv, Beto, Biruitorul, Bkonrad, Blethering Scot, Bob247, Bobrayner, BrownHairedGirl, CSWarren, Canada partner, Chris j wood, Claricia, Colonies Chris, Courcelles, Coyets, Crusoe8181, Cum Deo, Curtis3250692, DaL33T, Dardorosso, DaveOinSF, Davidships, Dolman50, Doron, Download, Durator, Dushan wideway, Ehrlich91, Eisbein1971, Enaidmawr, Essexman, Ev, ExcelsiorHH, FMB, Fratrep, Frisia Orientalis, Funandtrvl, GRuban, Getteenagekicks0, Gilliam, Gimboid13, Goran.S2, Haddison, Halsteadk, HandigeHarry, Hauzkhas, Hersfold, Ichbinauchda, Irvi Hyka, JB82, JLaTondre, JaGa, Joeb07, Jswfl09, Jza84, Kbthompson, Kelisi, Ken Gallager, KevinOB, Kleiner Tümmler, Knowing guy, Lars T., LeniDari, Leong0083, Libraire, Logan, Lotje, Lysy, MB-one, MRSC, Mai-Sachme, Marek69, Max von O, Memmingen, MetricSuperstar, MiG, Mick Knapton, Moleman2000, Musikus96, Mythsearcher, Naersjoen, Nick Number, Noclador, Ohconfucius, Olivier, Omnijohn, Onewayfare, Ospalh, Owain, Paxsimius, Pbl1998, Pilettes, Plehn, Psvaca, Pulv, R'n'B, RadioFan, Readro, Regan123, Rich Farmbrough, Rosh2610, Rossenglish, Saga City, Schlüggell, SchreiberBike, Sciencebloke, Scotthatton, Sebastian scha., Shaunj4, Silverhelm, Smjg, Spellmaster, Stone, T0pem0, Tabletop, Tassedethe, Tatasport, Tedernst, The C of E, Theoaktree, Thorad, Thoroe, Tone, Travelbird, Ulric1313, Veermaster, Vegaswikian, WOSlinker, Wefa, Welsh, Whouk, Woodshed, Woohookitty, Wosch21149, Wsswfrench, YUL89YYZ, Zé da Silva, 346 anonymous edits

Bavaria *Source*: http://en.wikipedia.org/w/index.php?title=Bavaria *Contributors*: (, -Ilhador-, 10 year old boy, 1000Faces, 111Alleskoenner, 128.196.133.xxx, 159753, 16@r, 2001:718:9:1:0:0:0:5, 2812, 2help, 52 Pickup, 842U, 87v7t76fc4iguwevf7657436253yd4fug754ws67dtfugiy67t8576, @xel, A history of the modern world, AAGUNMAN, ACV777, Abisch, Achates, Adam78, Adamk 97, Adolphus79, Adrian J. Hunter, Agathoclea, Agne27, Ahoerstemeier, Ahuffski10, Akkolon, Alarob, Aleenf1, Alekjds, Alensha, Alexander Domanda, AlexiusHoratius, AllPeopleUnite, Alvaro, Amatulic, Andi1995, AndrewRT, Andykentisi, Anger22, Angr, Animum, AnnaFrance, Anrod, Antandrus, Ardel Murzighiz, ArtVandelay13, Arthur Holland, Ask123, Astros4477, Attilios, AuburnPilot, Auslli, Avalon, AxSkov, BDD, BMWLover80, BYF, Baldhur, Baridiah, Bavarian Hiasl, Ben Ben, Ben-Zin, Bender235, Bermicourt, Betterart, BigMike718, Bigbossfarin, BilCat, Billinghurst, Bit swizzler, Blablablob, Bloodofox, Bloomfield, Bobblewik, Bobo192, Bogey97, Bomac, Boson, Brabban, Braincricket, Brichtig, Brion VIBBER, BsBsBs, Buaidh, Burzmali, C.Kent87, Cacetudo, CalicoCatLover, Calistemon, Caltas, CambridgeBayWeather, Canyouhearmenow, Caponer, Cdecoro, Cfaerber, Chanheigeorge, Charles01, CharlotteWebb, Chd, Chekaz, Choster, CieloEstrellado, Cinnamon42, Ckatz, ClickRick, Cloachland, Closedmouth, Cntras, Coemgenus, Colonel Cow, CommonsDelinker, Conscious, Conversion script, CopperSquare, Cosal, Coughinink, Craigzomack, DTOx, DWC LR, Dagbert, Dale Arnett, Dandelions, DanielCD, Darren Jowalsen, Darth Cookie Monster, Datu Dong, Dbachmann, Dbcook256, Dcandeto, Dcb1995, Delta G, Demcs, Den fjättrade ankan, Denelson83, Denihilonihil, Der Eberswalder, Derekbd, Devoblue, Dewritech, Discospinster, Djchris1991, Djmutex, DocendoDiscimus, Doco, Docu, Domino theory, Donama, Drindl, Drunken Pirate, Dysepsion, ESkog, East718, Easter Monkey, Edcolins, Edward NZ, Einemnet, Ejosse1, Ekwos, El C, Elassint, Elockid, Elox, Energyfreezer, Epbr123, Erianna, Erichwiki, Erikeltic, Euchiasmus, Evadb, EverettColdwell, Fabartus, Fabianklenk, FactStraight, Fallschirmjäger, Fantasy, FastLizard4, Favonian, Felix.R.A., Ferkelparade, Fingers-of-Pyrex, Finnrind, Fishwristwatch, Fluffernutter, Fogeltje, FordPrefect42, Fractalizator, Frappyjohn, Frietjes, Fuddle, Fulcher, Future Perfect at Sunrise, Fxhomie, Fyyer, GCarty, GJR, GLGermann, Gadget850, Gaius Cornelius, Gamma, Gargolla, Garybel, Garzo, Gavia immer, Gecko78, General Jazza, Gentgeen, Ghosts&empties, Gilgamesh, Gilgamesh he, Gilliam, Giraffedata, Glane23, Gnu1742, Gogafax, Gogo Dodo, Goldfishbutt, Gppande, Gr1st, Graham87, Grandiose, Grenavitar, Grimhelm, Grstain, Gryffindor, Gschmarri, Gurch, Gwernol, H005, Hackercraft, Hadal, Hairy Dude, Hajenso, Halsteadk, Hanberke, HansWorst, Hardman1, Harrystein, Hashar, Hayden120, HeartofaDog, Henrygb, Hibernian, Highpriority, Hingster, Hinicome, Hixteilchen, Hmains, Hokanomono, Hoo man, Hornymanatee, Hristarhos, Hu12, Huntthetroll, Husnock, Hvn0413, Hydrargyrum, IEdML, IZAK, Iamiyouareyou, Ian Pitchford, Immanuel Giel, Imperator24, Int21h, Isis, Ithurts2poo, Ivirivi00, Ixfd64, J. 'mach' wust, J. Patrick Fischer, J.P.Rallizgard, J.delanoy, J187B, JM.Beaubourg, JaGa, Jack.Danish, Jamesfett, Jamie Mercer, Jan eissfeldt, Jared Preston, Jasonuhl, Jcouser, Jeannedeba, Jebus989, Jeffman52001, Jeronimo, JesseRafe, Jesster79, Jguad1, Jiang, Jj137, Jnestorius, Joewein, John K, John Vandenberg, JohnBlackburne, Jonas Liljeström, Jorunn, Joseph Solis in Australia, Joshuashearn, Jpgordon, Jpxt2000, Ju98 5, Juergen1, Justinc, Juxtaposit, Jyusin, Kait, Kate Phaye, Kato2k6, Kelisi, Kesac, Kgrad, Khajidha, Kingjeff, Kinneyboy90, Kirandinakaran, Kjamesh89, Kortnay08, KrakatoaKatie, Kristinae, Kritikos99, Krun, Ksademap, Kuru, Kusma, Kwamikagami, KyraVixen, LOL, Laca, Lacrimosus, Lamro, Lars T., Lear 21, Lemonade51, Leszek Jańczuk, Lethe, Lfh, Lightmouse, Lipedia, Lir, LorenzoB, Lswartz, Luatha, Lumpydog32, Lysy, MER-C, MONGO, MaGioZal, Mac c, MacTire02, Mackeriv, MadGeographer, Magnus Manske, Malerooster, Malleus Fatuorum, Mandarax, Manecke, Maproom, Maralex, Marcika, Marek69, Markofthebeast, Markus Kuhn, Markus Stamm, Markussep, Martarius, Martin451, Marykeiran, Matthias Klostermayr, Mattissa, Maximus Rex, Mcferran, Meaghan, Melancholia i, Merovingian, Mevsfotw, Michael Hardy, Michael.palomino-at-gmx.ch, MichaelTinkler, Michel rockstar, Mirage, Mirekmarut, MisterGugaruz, Misza13, Mjquinn id, Monkeyblue, Moocha, Moombasa, Moomoocowsareawesome, Moonriddengirl, Morevisit, Morwen, Mr.K., MrBill47, Msedwick, Mxn, MyPasswordIs-FreeState, Müdigkeit, Nach0king, Narayansg, Nauticashades, NawlinWiki, Neolandes, Net-net, Nick Number, Nickshanks, Nico, Nigel Campbell, NoPuzzleStranger, Novil Ariandis, Number1schumacherfan, Numbo3, Ocee, Olessi, Olivier, Omicronpersei8, Omnieiunium, Orangemarlin, Oreo Priest, OwenBlacker, Oxfordwang, P. S. F. Freitas, Paine Ellsworth, Pascal.Tesson, Patrick, Paulc206, Peeperman, Penwhale, Pepsitest, Peregrine981, Peter Isotalo, Peteweez, Phearson, Philip Trueman, Philopedia, Picapica, Piccolo Modificatore Laborioso, Pigpen, Pilotguy, Polozooza, Poosaviour93, Postcard Cathy, Prjaeger, Proofreader77, Prosbus, Pzich, Qarakesek, Quadell, Qualiabavariae, Qxz, R'n'B, R-41, RRKennison, Rabauz, Rabbitrunning2, Rahuloof, Ramckay, Rarelibra, Rast5, Ratzer, Ravidreams, RayquazaDialgaWeird2210, Rbb 1181, Rchamberlain, Rdb, Redsox00002, Reindra, RexNL, Reymma, Reywas92, Rhombus, Rich Farmbrough, Richie, Ringbang, Rivertorch, RjCan, Rjbkearney, RonPab, Ronaldtheroofer, Rosenknospe, Rosh2610, Rothorpe, Rrius, Ruhrjung, Rupertslander, Ruud64, RyanGerbil10, Ryulong, Rōnin, S3000, STB110, Saddhiyama, Saluton, Sam Blacketer, Sanchristobal76, Sandman, Santryl, Sarah, Sardanaphalus, Scarian, Schneelocke, SchuminWeb, SchweizerJunge, Scientizzle, Scriberius, Sd31415, Seatdistrict, SebastianHelm, Sejdb, Shanes, Shilkanni, SidP, Signalhead, SilkTork, SilverWiki, Silverhelm, Sion8, Skibbledee, Skysmith, SlaskiAnanasek91, Sleigh, SluggoOne, Slysplace, Sm8900, SmartGuy Old, Smith2006, SnapSnap, Snowdog, Snoyes, SoWhy, Sobebunny, Son sonson, SpartanPhalanx8588, SpookyMulder, Srikeit, Srose, Starranger00, Steel, Stemonitis, Str1977, Sunderland06, Supertouch, Sverdrup, Symposiarch, Synchronism, TKD, Tabletop, Tagishsimon, Techman224, Tedius Zanarukando, Template namespace initialisation script, Tesi1700, Testsieger73, Texfive, The Behnam, The Photon, The Rambling Man, The Transhumanist, TheKMan, Thecheesykid, Thedjatclubrock, Thehelpfulone, Themfromspace, Theo van der Zalm, Thisisanoying, Thunderboltz, Thylacoleo, TikiWiki, Timwi, Titoxd, Tobby72, Tommycw1, Tony1, Tpbradbury, Traceylily, Traff1997, Troglo, Ttownfeen, Tulkolahten, Twhite 88, Twp, Ulf Heinsohn, Ulritz, Unyoyega, Urban XII, Ussschrotti, UtherSRG, Uthlande, Valentinian, Valrith, Vanished User 1004, Vanjagenije, Vaquero100, Vdiest, Vdjj1960, Verazzano, Vertium, Viewfinder, VodkaMartini, Walor, West.andrew.g, Whouk, Whugel, Wi-king, Wilfried Derksen, Wilhelm meis, Winchelsea, Woohookitty, WordyGirl90, Writtenright, Wronggray2012, Wzwz, Yngvadottir, Yooden, Yossiea, Zebra848, Zeno Gantner, Ziko, ZingbatStew,

Zinnmann, Zundark, Zzzzzozzzzz, 1107 anonymous edits

Germany *Source*: http://en.wikipedia.org/w/index.php?title=Germany *Contributors*: -- April, -Majestic-, 02pollajo, 03md, 06alifar, 111Alleskönner, 119, 123kelsey123, 1297, 14Adrian, 159753, 16@r, 1Winston, 1exec1, 247balla34, 2812, 334a, 52 Pickup, 7&6=thirteen, 7huizen, 8ung3st, A Werewolf, A bit iffy, A little insignificant, A-giau, ABCD, ABlockedUser, AJackl, ALE!, ANNRC, Aaker, Aaron, Aaron Einstein, Aaron Schulz, Aaronlws, Aazn, Abe Lincoln, Abonazzi, Abrech, Academic Challenger, Academicigbo, Acalamari, Acanon, Achates, Adam Bishop, Adam Lutostański, Adashiel, Adcva, Adherent of the Enlightenment 10.0, AdmRose, Aeusoes1, Agathoclea, Aherunar, Ahoerstemeier, Ahrarara, Ahuskay, Aitias, Aivazovsky, AjaxSmack, Ajenterp, Akanemoto, Akhristov, Akkolon, Akradecki, Alai, Alan De Smet, Alanmak, Alarics, Alberto3166, Alex S, Alex Usoltsev, Alex1011, Alexandru Busa, Alexandru Stanoi, Alf54354353, Alfirin, Alfonsomedina1, Algebraist, Algont, Alison, Alixus, All N Ever, Allgermein, Allmightyduck, AllyUnion, AlmightyClam, Alopex, Alphachimp, Alphasinus, Alphathon, Alphax, Altenmann, Alterrabe, Amakuru, Amaury, Amber-W-12, Amcaja, AmiDaniel, Amoreno65, Anaki72, Ancheta Wis, Andre Engels, Andrew Gray, Andrew Levine, Andrewlp1991, Andries, Andrij Kursetsky, Andris, Andrwsc, Andy Marchbanks, Andyjsmith, Andypandy.UK, Anetode, Anger22, Angr, AniRaptor2001, Animum, Anita5192, Anjwalker, Ann O'nyme, Anna Lincoln, Anoldtreeok, AnonMoos, Anonymous from the 21th century, Antandrus, Anteeru, Anthony3, Anthonymendoza, Antidote, Antientropic, Anton88be, Antonio Basto, Ap, Apeloverage, Apoc2400, Apokrif, Appleseed, Apuman, Aqwis, Arathjp, Arbero, Arindam.2011, Aris Katsaris, ArmadilloFromHell, Armogamer13, Arnomane, Arnowaschk, Arsonal, Art LaPella, Artichoker, Arunnplus, Asams10, Asarelah, Asdfasdf1231234, Ashton1983, Askalan, Asnatu wiki, Asocall, AsusFan, Ataraxis1492, Athinaios, Aths, Atlant, Atlanticpuffin, Atlantik, Atoric, Atrix20, Attilios, Atwardow, AuburnPilot, Aude, Audiovideo, Aufbauten, Augustus Rookwood, Auntof6, Auric, Auriong, Autocracy, Avala, Avatarion, AvicAWB, Avicennasis, Avraham, Awolf002, AxG, Axegod12, Axel Phony, AxelBoldt, Axpde, Axt, Aye Carumba Fajita Pizza, Azate, Aznpryde, B1llycl3m3nt, BD2412, BGManofID, BJess, BRG, Babakexorramdin, Babbage, Backspace, Baldhur, Ballsontheroad, BalowStar, Balph Eubank, Bambuway, Banana04131, BanyanTree, Bapon, Barberio, Bardhylius, Barek, Baristarim, Barliner, Baronnet, Barras, Barryob, Barutazaru, Baseballnut290, Bastin, Battem, Baudrillard, Bazonka, Bbenjoe, Bbruchs, Bcorr, Beagel, Beetstra, Bejinhan, Beland, BenMerill, Bender235, Benea, Benefitov, Benesch, BenjaT, Benne, Benster1510, Bento00, Benwildeboer, Benznine, Berkut, BernardaAlba, Bernardissimo, BernhardtP, Bert-25, Beta m, Betacommand, Betterworld, Bevo74, Bg007, Bhuck, BigBen212, Biker Biker, BilCat, Bilbo571, Bippy1990, BitterSTAR, Bjarki S, Bkell, Blablaaa, Blacha, BlackGothFaerie, Blackjack48, Blacklite, Blake24, Blake3522, BlazeTheMovieFan, Bleh999, Bletch, Blinder Seher, Blizzardstep0, Bloodofox, Blue-Haired Lawyer, Blue520, BlueMars, Bluemask, Bluezy, Blur4760, Bml, Bob rulz, Bob5409, Bob889900, Bobblewik, Bobby122, Bobet, Bobo192, Bocafan76, Bohuiginn, Boothy443, BorgHunter, Bornsommer, BoroughPeter, Boson, Bowman17, Boznia, Brad101, Bradeos Graphon, Brahmaputra, Brain40, Brainbark, Branddobbe, Brandmeister, Brandmeister (old), Brandt1, Breathe, BrendelSignature, Breno, Brethrickacey, Brian0918, Brighterorange, Brion VIBBER, Britannicus, BritishWatcher, Brockert, BrokenSegue, Brossow, Bruceknibbs, Brumski, Brutaldeluxe, Brutannica, Bryan Derksen, Bssc81, Buaidh, Bubba hotep, Buchanan-Hermit, Buddha24, BumHoleo, Bunny-chan, Bushmillsmccallan, Bushy moustache, Butwhatdoiknow, BuzzWoof, Bwithh, ByrnedHead, CJ Aikman, CORNELIUSSEON, CSWarren, Cactus.man, Caerwine, Cakeandicecream, CalJW, CalendarWatcher, CalicoCatLover, Caltas, Calum MacÙisdean, Cam275, CambridgeBayWeather, Cambyses, Cameltoe237120236419324, Camilton, Can't sleep, clown will eat me, CanDo, Canadian-Bacon, Canaen, Canationalist, Canderra, Caniago, Canterbury Tail, Cantus, Captain Courageous, Captain Future, CaptainUnderpants, CaptainVindaloo, Captainspizzo, Carabinieri, Caratinga8, Carifio24, CarlCarlton, CarlMosel, Carlossuarez46, Carnby, Carson56437, Caryptes, Casper2k3, Catalyst in Society, Catdude, Causa sui, Cause of death, Cautious, Caveman 07, Cedrus-Libani, Celestra, Celyndel, Cenarium, Censusdata, Centrx, Cethegus, Ch'marr, ChKa, Chaddy, Chairboy, Chalst, Chanting Fox, Chaosdruid, CharlesMartel, Charly33, CharonX, Chatanga, Che829, Checkmate987, Chederman69, Cheeseheadxvi, Chicheley, Childzy, Chillum, Chipmunkdavis, Chochopk, Chodorkovskiy, Chokerman88, Choop, Choster, Chowbok, Chris 73, Chris Archer, Chris the speller, Chrisfow, Chrisritch, Christian List, Christinam, Christoph-NZ, Chriswiki, Chrysalis, Chtrede, Chubbles, Chuck Carroll, Chucky21, Church of emacs, Chuunen Baka, Chwyatt, Chzz, Ciacchi, Ciceronl, Ciddler, CieloEstrellado, Cirt, Citizen-of-wiki, City-17, Civil Engineer III, CjDMaX, Cjthellama, Ckatz, Ckorff, Classical 2006, Cliché Online, Cloyd11, Cmathio, Cmdrjameson, Coasterlover1994, Coat of Arms, Codycod10, Coffeinfreak, Cogiati, Cogito ergo sumo, Cognate247, Cokes360, Colchicum, ColdWind, Colonies Chris, CommonsDelinker, ConWilBur, Conor's Whipped, Conorbrady.ie, Conorobradaigh, Conte di Cavour, Conti, Contributor777, Conversion script, Cooksey, Coolcole93, Cordless Larry, Coredesat, Cosmic Latte, Crabbit, Craigzomack, Crazycomputers, Crazycoolalex, CrimsonBlue, Crisskriss, Crissov, Crito2161, CrnaGora, Crossk, Crowbar1234, Crusadeonilliteracy, Cruzian, Cryptic, Crystallina, Crzrussian, Cs-wolves, Csörföly D, Cukiger, Culveyhouse, Curps, Curtisvic, Cvieg, Cyde, Cyfal, Cyopardi, Cyrius, D0rkypn0y1029, D34d10gic, D6, DARTH SIDIOUS 2, DCGeist, DGJM, DIREKTOR, DJ Clayworth, DLiebisch, DMacks, DO'Neil, DTGHYUKLPOQWMNB, DTOx, DVD R W, DaL33T, DaQuirin, Dakhart, Dalmatian Mommy, Damian Radu, Dan D. Ric, Dan Guan, Dan100, DanMS, Dana boomer, Danaman5, Danc, DancingPenguin, Daniel5127, DanielCD, Danielt, Danny, Dantheman531, Dapete, Daqui, Darealclub, DarkFalls, Darkieboy236, Dasani, Dave031970, Dave420, Daveb, Davesmall, David J. Rogers, David Kernow, David Levy, David Liuzzo, David N. Hermann, David.Mestel, Davidfollsom, Davidpdx, Dbachmann, Dca5347, Dcandeto, Dcflyer, Deadblob93, Deanos, Debate789, Debresser, Dedeler31, Deejaye6, Deflective, DefunKt, Deiz, Delldot, Deltabeignet, Demiurge, Democryt, Demomoke, Dendodge, DennisDaniels, DennisGigio, Dennisleckwee, Denniss, Deposuit, Deppp, Der Eberswalder, Der Statistiker, Der Wolf im Wald, DerHexer, Deszcz, Deutsche, DevastatorIIC, Devatipan, Deviathan, Deville, Dewritech, Dfrg.msc, Dgg32, Dhp1080, DickLukensTheSecond, Dierk, Dinner123, DirectorR, Directorstratton, DirkvdM, Dirtyhose, DisappointedHorse, Discospinster, Dispenser, DivineIntervention, Dizzyizzy, Djmutex, Dkusic1, Dlugopis, DocendoDiscimus, Doco, Doctor Allen, Docu, Dogface, Dojarca, Donnamaria, Donnerstag, DopefishJustin, Dorftrottel, Doric Loon, Doubleodd, Doug Johnson, Dowj, Downing Street, Dpr, Dqeswn, Dr. Crisp, Dr.alf, DrBob, DrKiernan, Dragoburaggo, Dragoon009, Drewbean, Drewrau, Drgals, Driftwood87, Drmies, Drunkenmonkey, Dub4u, Dude8679, Dukeofomnium, Duncharris, Durin, Dusty relic, DutchmanJoy, Dysmorodrepanis, Dzenanz, E Pluribus Anthony, E-boy, E235, EBB, ESkog, Easter Monkey, Eatyourgreens, Eco84, Eddospants, Eden Tate, Edgar181, Editor at Large, Editor18, Edwy, EffK, Ehrenkater, Eirik.lingas, Eixo, Ejgreen77, Ekki01, Eklipse, El C, Electionworld, Elekhh, Eleuther, Elf-friend, Elfguy, Elizabeth A, Elliskev, Ellmist, Elmondo21st, Elockid, Eloquence, Elvis, Emperorbma, EncycloPetey, Endofskull, Endurance, EnemyOfTheState, Energyfreezer, Engleaugen6666, Englishnerd, Enlil Ninlil, Enviroboy, EoGuy, Eon, Er Komandante, Eraliswelsh, Erauch, Eric Yurken, EricS, Ericoides, Erlando, ErnieBern, Ernio48, Erstats, Erudra, Esanchez7587, Escape Artist Swyer, EscapingLife, Eschbaumer, Esimal, Etams, Eu.stefan, Euchiasmus, Eugen Simion 14, Eugene van der Pijll, Everyking, Evice, Evil Monkey, Ewpfpod, F2020, FAThomssen, FF2010, Fabartus, Factuarius, Fahima07, Fainites, Fair Deal, Fallschirmjäger, Fama21, FamilyGuy1998, Familyrits, Farooqmoazzam, Fastestdogever, FatM1ke, Favonian, Fawhad, Feldfrei, Feldmarschall, Felipito1.966, Felixboy, Ferrija1, Ferro Carlotta Monzi Brak, Feuerspiegel, Fheidener, Fhumeres, Fidelio72, Fiet Nam, Filelakeshoe, Finalius, Finlay, FinnishDriver, Firetrap9254, Firsfron, Fisss, Fkbreitl, Flamdring, Flammingo, Flappychappy, Flatterworld, Flauto Dolce, Fleris, Florentino floro, Florian Blaschke, Flowanda, Flowerpotman, Fnfd, Foncea, Fonzy, Fooloomanzoo, ForOurXylophones, FordPrkt, Forstnera, Fossa, Fotaun, Four-ArmedRevenge, Fourtimesfour, Franciswiki, Francs2000, Frank Anchor, Freak in the bunnysuit, Freakincoursework, Freakofnurture, Fred Stober, Freddo207, FreplySpang, Friedrichshainer, Frietjes, FritzK0310, Frymaster, Fschoenm, Fsotrain09, Fudge43, Fudoreaper, Fujishima, Fulcher, Funkelnagelneu, Funnyhat, Furrykef, Future Perfect at Sunrise, Fuzheado, Fuzzy510, Fvincent, Fvw, Fwappler, Fyodor, G-Man, G. Campbell, GBWallenstein, GGT, GHe, GLGerman, Gabbe, Gabithefirst, Gabriel-Royce, Gadfium, Gaius Cornelius, Gaius Octavius Princeps, Galati, Galoubet, Gamer007, Garryq, Gausie jr, Gavia immer, Gavin Moodie, Gaz, Gekritzl, GenQuest, Geneb1955, Geni, Gennarous, Geo-Loge, Georg Hurtig, GeorgeTSLC, GeraldH, Gerhard51, German Food, German name, GermanJoe, Germanclub, Germanuser, Gert7, Gertjan R., Gertyness, Getker, Gfoley4, Gggh, Ggoere, Ghepeu, GiZiBoNG, Gidonb, Gilbertfein, Gilgamesh, Gilliam, Gilo1969, Gimmetrow, Giovanni76, Giraffedata, Girdi, Gironauni, Giving is good, as long as you're getting, Gjp23, Glen, Glenn, Gmaxwell, Gmosoti, GnuDoyng, Godefroy, GoetheFromm, Gogafax, Gogo Dodo, Golbez, Good Olfactory, GoodSirJava, Goodoldpolonius2, Googl, Googleguy1234, Gotrl, Gr8opinionater, Grace Note, Gracenotes, Grade7, GraemeL, Grafen, Grafikm fr, Graham87, Gran2, Graufbold, Greatal386, Greatgreenwhale, GreekNurse, Green Giant, Green Politics, Green caterpillar, GreenUpGreenOut, GregAsche, Grendelkhan, Greyhood, Griffinofwales, GrimmC, Grinman, Grinq, Grmpf, Gronau, Grstain, Grunt, Gryffindor, Gsklee, Gtg204y, Guanaco, Gugganij, Guidedbyalan, Guideended1, Guilherme Paula, Guliolopez, Gunnar.Kaestle, Guoguo12, Gurch, Gurchzilla, Guy M, Gwernol, Gwyon, Gzornenplatz, H-stt, H0ttamale2, H1523702, HIDECCHI001, HJ Mitchell, HVH, Hadal, Hadi Payami, Hadseys, HagSibylle, Haham hanuka, Hairy Dude, Hajduk, Hajor, Half price, HalfShadow, Halibutt, Hall Monitor, HamboLSSAH, HammerFilmFan, Hannil, HappyCamper, Harald Meier, Harmil, Harpsichord246, Harryboyles, Hasan Dalbudak, Hassocks5489, Hatt, Haukurth, Hawkestone, Hayden120, Haymaker, Hazhk, Hazzl, Hdante, Heah, Heimdal, Heimstern, Hektor, Helsinkicommission, Helwerda, Hemanshu, Henry Flower, HerkusMonte, Heron, Herostratus, Herr Kent, Herra660, Hesus321, Hexil blitz, Hhielscher, Highfields, Highpoems123, Hishii, Historian19, Hmains, Hmwith, Hoc30, Holonboy, Holy Oil And Water, Holynightfever, Homagetocatalonia, Homerjay, Honette, Hoo man, HoodedMan, Hornymanatee, Horst-schlaemma, Househelps, Hqb, Htews, Hugo999, HungarianWoman, Husamelodeh, Husond, Hydrargyrum, Hydraton31, I already forgot, IIIraute, IJA, IJzeren Jan, Ian Pitchford, Ibn Battuta, IceArchon, Icecradle, Ich, Ich-AG, Ich901, Idaltu, Ideletealot2, Ief, Igo4U, Iksisy, Ilehmann, Ilgom, Illythr, Iluvcapra, Ilya1166, Imadcow1, Imaglang, Imlepid, Imperator, ImperatorExercitus, Imperium Europeum, Improv, Imtrogdor, Imwivstuipid, InShaneee, Indon, Infamouskitty, Information-Line, Inge-Lyubov, Ingebjorg, Ingolfson, Instantnood, Int21h, Inwind, Iol, Iowausa, Ipsofactoid, Irdepesca572, Iridescent, Irishguy, Iritebs222, IronGargoyle, Irrypride, IsarSteve, Istabraq, It Is Me Here, Italiano111, Ivirivi00, Ixfd64, Izehar, Izzedine, J Di, J Milburn, J. 'mach' wust, J.K Rowling, J.delanoy, J00tel, JAU123, JCAla, JFT96, JForget, JHMM13, JJARichardson, JLogan, JM.Beaubourg, JMS123, JNW, JTBX, Jacek Kendysz, Jack Silver1000, Jackobyte, Jacques2, Jadger, Jaellee, Jaggsan, Jajafe, Jakobburkhardt, James086, James297, Jamesooders, Jamyskis, Jan Hofmann, Janneman, Jansch, Jared Preston, Jarjarbinks10, Jarkeld, Jasman, Jatinbhatt blap, Jatkins, Java13690, Java7837, Jaxl, Jaxsonjo, Jayden54, Jayjg, Jazz+, Jbattersby, Jbou, Jcornelius, Jcrook1987, Jcw69, JdH, JdeJ, JeLuF, Jean.artegui, Jeff G., Jeff3000, Jeffrey O. Gustafson, Jekki, Jeltz, Jensboot, JeremyA, Jeronimo, Jerry67, Jessejamesandloop, Jessica Gordon, Jesushaces, Jesusjonez, Jez, Jfruh, Jhendin, Jiang, Jim bob bryar, Jimiloughlin, Jimp, Jjhcap99, Jjk82, Jklin, Jknabe, Jmcc150, Jmh, JoB614, Joachim Weckermann, JoanneB, JodyB, Joe Canuck, Johan Magnus, Johann Wolfgang, Johannes H.-Larsen, Johanneswilm, John, John D., John K, John R Murray, John Riemann Soong, John Smith M.D., Ph.D, John Smith's, John-vogel, John254, Johnny Undead, Johnpro, Jomika, JonGwynne, JonHarder, JonRoma, Jonah474747, Jonathan Groß, Jondel, Jondumitru, Jonesy, Jonophan, Jooler, Jor, Jose77, Joseph Solis in Australia, Joshua, Jossi, Joséwell, Joy, Jpeob, Jpgordon, Jr724, Jrcrin001, Jrleighton, Jrtsstfjsfkyjsrt6istk7stkyskjytrsk, Jsbhavsar, Jsc83, JuJube, Jujutacular, Julius1990, Jumbuck, Jumpytoo, Junes, Junkietom08, Jusdafax, Jusjih, JustPhil, Justin Eiler, Justin Piga, Justinc, Justinhful, Jwestbrook, Jwordsmith, Jwy, Jyusin, Jza84, KFP, KG200, KJohansson, KMileSupernova, Kacen, Kadillia, Kafziel, KagamiNoMiko, Kahusi, Kaihsu, Kaiser Torikka, Kaiserjd, Kajzderski, Kakofonous, Kalaha, Kamyra, Kantianer, Karam.Anthony.K, Karl, Karl-Henner, KarlFrei, KarlMathiessen, Katherz, KathrynLybarger, Katzensprung, Kbdank71, Kchishol1970, Keegan, Keesiewonder, Keichwa, Keilana, Keith D, Keizuko, Kejaja, Kelly Martin, Kelnor, Kermecke, Kevin, KevinCarmody, KevinOB, KevinTR, Kevs, Kevtrice, Kewp, Kewpid, Khanada, Khoikhoi, Khukri, Killdevil, Killer.k09, KillerChihuahua, Kim-Zhang-Hong, King of Hearts, Kiwifruit89, Kkm010, Klaus Bells, Klausness, Kleuske, Klosterdev, Kman543210, Kman665, Knarf-bz, Knipptang, KnowledgeOfSelf, Knowledgebycoop, Koavf, Komet, Kompakt, Konstantin, Koro lord, Kosebamse, Kotniski, Koyaanis Qatsi, Kozuch, Kr1st1deejay97, KrakatoaKatie, Krankman, Kransky, Krazytea, Krellis, Kribbeh, Kross, Ksenon, Kt66, Kungfuadam, Kupraios, KuroiShiroi, Kuru, Kusma, Kwamikagami, Kwesiidun91, Kwiki, Kyle Benge, Kyle4950, Kyrin, Körnerbrötchen, Kızıl Şaman, L'Etat c'est moi, L1A1 FAL, L95slovenia, LARS, LIU, LKDD, LLLex, LSLM, LaMenta3, Laca, Lacrimosus, LadnaV, Lafuzion, Lakshmix, Lan360, Lancevortex, Landroo, Langston, Larko3344, Larry Sanger, Larry_Sanger, Lars T., Lars Washington, LarsJanZeeuwRules, Latristelagrima, Laugreve, Laurens-af, Laurentleap, Lawl1232323, Le Fou, Leafyplant, Leandrod, Lear 21, Learngermanwords, Lectonar, Lee J Haywood, Lee S. Svoboda, Leewilson123, Leibniz, Lemsi, Lemsus, Leo Bauer, LeoO3, Lethe, Leuko, Levalley, Levdave7381, Levineps, Levonscott, Lexicon, Lfh, Lhademmor, Lhasaq, Liamdaly620, Liberatus, Libyan10, Lietzow, Liface, LightSpectra, Lightdarkness, Lightmouse, Ligulem, LilHelpa, Lilac Soul, Lilshimon, Lilybaeum, Linuxbeak, Lir, ListedRenegade, LiterallySimon, Lithpiperpilot, LittleDan, LittleOldMe old, LittleRoughRhinestone, LlywelynII, Llywrch, Lobsterantlers, Locke Cole, Lockesdonkey, Logan, Lolhilol, Lolitsjoel16, Lolol123, LonelyMarble, Lonerville, Longbow4u, Longhair, Lord

Vader, LordofHavoc, Lordvaluemart, Loren36, Loshaisamazing, Lotje, Louis3ham, Lowellian, Lrobb95, Lstipakov, Lucas323, Lucyin, Ludger1961, Ludjer, Lugnad, LuigiManiac, Luis rib, Luisztdt, Lukatheist, Luna Santin, Lunacy.Fringe, Lunar2, Lunarboy, Lupo, Lurker, Lyfeizshrt, Lysy, MAXXX-309, MECU, MER-C, MIKElovesgermanchicks, MJCdetroit, MKers06, MKoltnow, MOMBO BAGGINS, MONGO, MPS, MPerel, MTWEmperor, MaGioZal, MacedonianBoy, Macintosh User, Mackensen, Mackeriv, Mactruth, Madman, Maelnuneb, Magicjigpipe, Magioladitis, Magister Mathematicae, Magnus Manske, Magumbaboy, Mailer diablo, Malarious, Malbi, Malcolm Farmer, Male1979, Malerin, Malfidus, Malo, Man vyi, Mandavi, Manecke, Maneus4, Mangoe, Manop, Manpyo, MapsMan, Marcika, Marco polo, Marcus Cyron, Marcworca, Marek69, MarekF, Maria Stella, Marielleh, Mark, Mark.P.Bartlett, MarkThomas, Markus Schmaus, Marky-Son, Marlight, Marnanel, Marriex, MarsRover, Martg76, Martin.schoenberg, Martinp23, Martinwilke1980, Maru4u, Marvan, Marysunshine, Marzoregulus@yahoo.co.ph, Marzter6, Masken, Masonshowalter, Massjit, Master Jay, MasterLycidas, Materialscientist, MathFacts, Mathew123, MathiasFox, Matilda, Matros, Matstone99, Matt 989, Matt Crypto, Matt512, Mattbaileyisneat, Matthead, Matthew Fennell, Matthewjw 16, Matthäus Wander, Matticus78, Mattingly23, MatttK, Mauriziotani, Mav, Maverick starstrider, Mawa, MaxBech1975, Maximus Rex, Maxschmelling, Maxymax66, Maymose, Mayonnaise, Mayumashu, Mazzelino, McDogm, McGeddon, Mcewan, Mclay1, Mcmoor, Mcurmamma, Me Mighty Leper, Meelar, MegA, Meh31, Mehmeda, MeinGoshen, Melaen, Mellum, MementoVivere, Menameischarle, Menchi, Mentifisto, Merbabu, Merchbow, Mercury, Merlincooper, Mertan, Mervyn, Mesgul82, Metaljaws, Mfa fariz, Mgiganteus1, Mgoodyear, MiLo28, Miaers, Mic, Mich112358, Michael Hardy, Michael Zimmermann, Michael riber jorgensen, Michael.palomino-at-gmx.ch, MichaelDiederich, Michaelas10, Michaelmcneil, Michaeltobbylee, Michbich, Michealkelly, Michel rockstar, Mickey gfss2007, Microsomen, Middleeast, Midnightblueowl, Mifter, Mightymights, Miguel Andrade, Miguel.mateo, Mike Rosoft, Mike.heckmann, MikeAtari, Mikeifc123, MillaSukel, Milton Stanley, Mipago, Miroslawa, MisfitToys, Miskin, Missspencer06, Mister C4, MisterSheik, Misza13, Mittermeier, Mixcoatl, Miyokan, Mjpieters, Mmunger, Mnemeson, Mnmazur, Mo-Al, Mo0, Modelun88, Moe Epsilon, Moehre, Mollemy, Molobo, Moncrief, Monkeyknife, Monomachus, Moominjosh, Moonunit02, Mooseberry, Moriori, Morse thor, Morwen, Moshe Constantine Hassan Al-Silverburg, Mouvement, Movementarian, Moxdude, Moxy, Mqduck, Mr Adequate, Mr MillA, Mr Stephen, Mr Woman, Mr. Know-It-All, Mr. bwn, MrCanadian1, MrFish, MrPMonday, MrRandomPerson, MrTranscript, Mramz88, Mrfathead, Mrjeff, Mschel, Mshecket, Mtaylor848, Mtl 16, Mtmoore321, Mucky Duck, Mufka, Muhandes, Munci, Mus Musculus, Muskydusky, Mxcatania, Mxn, My76Strat, Myanw, Myscrnnm, N-HH, N-edits, N5iln, NB-NB, NYSTEX1992, Naive cynic, Nakon, NameIsRon, Narya, Naryathegreat, Nascar1996, Natalie Erin, NathanaelKing, Naverno, NawlinWiki, Nbatra, Nebauer, Nectarflowed, Nellov5, Neozoon, Nero the second, NeroN BG, NetherlandishYankee, Netknowle, Netsnipe, Neurowin, Neutrality, Nevfennas, NewYork483, Newkai, Newnoise, Nicander, Nichalp, Nick San, Nicketje666, Nickinny, Nickispeaki, Nickj, Nicklovesgold, Nicksukh, Nicktenoshi, Nico, Nicofs, Nicolas1995, Nieding, NigelR, Night w, Nihil novi, Nika, Nikkimaria, Niko226, Nikola Smolenski, Nikostar, Nimbusania, Nirvana888, Nishkid64, Nitya Dharma, Nivix, Nixdorf, Njecoolguy90, Nlnnet, Nmpenguin, NoGringo, NoPuzzleStranger, Nobleeagle, Nohomers48, Nolelover, Nonpawit, Noon, Noozgroop, Nopetro, Northernmonkey, Nothlit, Notuncurious, Nouse4aname, Npovshark, Nricardo, Nscheffey, Ntwarrior, NuclearVacuum, NuclearWarfare, Nufy8, Numbo3, Nunquam Dormio, Nv8200p, Nzpcmad, OOODDD, OSX, Obarskyr, Octane, Od Mishehu, OettingerCroat, Ohconfucius, Ohnoitsjamie, Okichubiness, Okinawadude, Olaf Eriksson, Old Guard, Oldnag85, Olegwiki, Olessi, Olgaru, OliFer, OliverH, OlliffeObscurity, Olorin28, Olowek, Oman66, Omicronpersei8, Onco p53, OneGuy, Oneearth, Onlineinteraction, Oo64eva, Oosoom, Opc123, Opelio, Opus88888, Orangerider, Orangutan89, OrbitOne, Oreo Priest, Ori.livneh, Ornithologician, Orz, Ospalh, Ot, Ottawa4ever, Otto ter Haar, OwenX, P-Schmidt22, P.juka, PBS, PFHLai, PS2pcGAMER, PSWG1920, Paddyez, Paine Ellsworth, Paintball12345, Pakhomovru, Pakon111, Palica, Pan Dan, Panoramastitcher, Pantherskin, Paradocks2, Paranoid, Parkwells, Pascal.Tesson, Pathless, Patka, Patrick, Patstuart, Patyk, Paul Benjamin Austin, Paul Erik, PaulBommel, Pavlo Chemist, Pawel z Niepolomic, Pax:Vobiscum, Peacock28, Pearle, Pecher, Penmachine, Pensionero, Pepsidrinka, Per Abrahamsen, Peter, Peter Horn, Peter Isotalo, Peter2005, PeterCS, Peterale4444, Pethr, Petree13, Petri Krohn, Petrux, Pewwer42, Pgk, Phaedriel, Phantomsteve, Pharaoh of the Wizards, Phatom87, Phenz, Phil Bastian, Phil Boswell, PhilKnight, Philip Cross, Philip ea, Phillip J, Philophobia, Philter137, Phoenix of9, Picapica, Picaroon, Piccadilly, Piggyspizza, Pilotguy, Pinkville, Piotrus, Piratelooksat30some2001, Pirawik, Pjacobi, Platane, Plocho22, Plumber, Poarps, Pofka, Pokrajac, Polaron, Polylepsis, Pommerland, Popadopolis, Porqin, Porsche997SBS, Postcard Cathy, Postdlf, Postmann Michael, Prattlement, Preslethe, PrimeCupEevee, Pristino, ProhibitOnions, Proteus, Prumpf, PsY.cHo, Pschemp, Pseudo-Richard, PseudoSudo, Psg116, Pslide, Psy guy, Puchiko, Punctilius, Purple Sheep, Puttputttime, PuzzletChung, Pyroblondie, Qelp, Qero, Quakenul, QuartierLatin1968, QueenStupid, Queson, Quibik, Quite vivid blur, Quizimodo, Quizimodo (usurped), Qweedc, Qwerasdfzxcv1234, Qwerty Binary, R Lowry, R'n'B, R-41, R9tgokunks, RG2, RHB, RM21, RMHED, RPlunk, RSido, Raamin, Rachel5D, Rada, Radagast, Radu Gherasim, RaffiKojian, RagingR2, Railer 353, Ralphy, Ran, Randomblue, RandySavageFTW, RapaNui75, Rarelibra, Raul14, Raul654, Raven4x4x, Ray Van De Walker, Raymond arritt, Rbz321, Rd232, Reboot, Reconsider the static, Recury, Red Winged Duck, RedWolf, Reddi, Redking7, Rednaxela, Reedy, Reimanova, Reindra, Reinis, Reisio, Rejedef, Renata, ResearchRave, Resilienzi, Retired username, Retireduser1111, Rettetast, Revolving Bugbear, Rex Germanus, RexNL, Reywas92, Rflesher, Rhys fry, Riana, Ribbet32, Ricbep, Rich Farmbrough, Richard D. LeCour, Richardcavell, Richarddd, Richardfabi, Rick Block, RickK, Ricky@36, Rickyrab, Rilekou, Rimshot, Rje, Rjensen, Rjwilmsi, Rl, Rmky87, Rmt2m, Rnelson, RoMex, RoadTrain, Rob 06-07, Rob.derosa, Robdurbar, Robendo, Robert1947, Robert25, RobertG, Robertsteadman, Robinstocks, Robomaeyhem, Robzz, Rocastelo, RockMFR, Rocket000, Rockfang, Rockstar4000, Rod57, Rogerd, Rokbas, Romankawe, Romanm, Ronhjones, Rosenzweig, Routerrandy4, Roxi2, Roy da Vinci, RoyBoy, Royboycrashfan, Rracecarr, Rreagan007, Rsquire3, Rtc, RuffRyderMohan, Ruhrjung, Rururuu, RussBlau, Ruud64, RxS, RyanGerbil10, Ryulong, SFC9394, SKC, SNIyer12, Sacxpert, Saforrest, Sagaciousuk, Said 13, Saimdusan, Saintswithin, Salion, SallyForth123, Salmnj, Salsa Shark, Salsb, Sam Korn, Sam Spade, Sam Vimes, Sam-m, Sam01257462071, SamEV, Samantha555, Samashton, Samgardner94, Samir, Samrolken, Samsara, Samuel, Samuelsen, Sandersonjoe221, Sandman, Sandpiper, Sandwriter, SandyGeorgia, Sango123, Sannse, Sarcelles, Sardanaphalus, Satori, Savidan, Saxxtreme, Sca, Scartboy, Sceptre, Schaeferchr11, Schaengel89, Scheibenzahl, Schlumpff, Schneelocke, Schultz.Ryan, Schumacherfan, SchuminWeb, Schwatdy, Schzmo, Scifiintel, Scipius, Sciptor, Sciurinæ, Scohoust, Scott777, Scovetta, Scrollriter, Scryer 360, Scythian1, Seaman, Sean William, Seb az86556, Sebesta, Secret (renamed), Section6, SeeSchloss, Selinger, Semifriend, Semprini, Separa, Serafin, SergioGeorgini, Serpicotv, Serte, Sevela.p, Sfahey, Sfan00 IMG, Sgeureka, Sgrayban, Sgt Simpson, Shadiac, Shadow007, Shadow1, Shadowjams, Shamir1, Shanel, Shanes, Shannon bohle, Shastrix, Shenhemu, Sheogorath, Sherollin, Sherool, Shivan, Shlomke, Shoaler, Short Verses, Shredder2012, Shreshth91, Shwangtianyuan, Sicilianmandolin, SigPig, Signsolid, Simetrical, SimonP, Simonides, Sintaku, Sir Isaac, Sirius85, Sjacinth, Sjakkalle, Sketchmoose, Skinnyweed, Skizzik, Sky, Sky Diva, Skydeepblue, Skysmith, Skywolf, Skäpperöd, SlamDiego, Sleigh, Sleptrip, Slomox, Slowking Man, SluggoOne, Smaalouf, Smith2006, SmthManly, Smug Irony, Smyth, Snack417, Snagari, Snazzcen, Snickers45, Snieg, SnowFire, Snowolf, Snoyes, SoCalSuperEagle, SoLando, SoWhy, Softjuice, Solicitr, SolidEdgeGuRu, Solipsist, SomeHuman, Something Wicked, Somethingoranother, Son of Al, Soriamic, Sortior, Sousclef, Soxwon, Sp33dyphil, SpNeo, Space Cadet, SpamBilly, Specialplant, Spitfire, Spitfire19, Spitzl, Splodgeness, Spongefrog, SpookyMulder, Springeragh, Springwalk, Spyro02, Squalie21, SqueakBox, Squirrelofdoom5, Srich32977, Srndp8e, Ssault, Ssolbergj, Staatenloser, Stallions2010, Stangaa, Stealth012, Steel, Stefan Udrea, Stemonitis, Stephan Schulz, Stephen G. Brown, Stephenb, Stephreef, Stern, Stestagg, Stetay, Stevemangold, Stevencool104, Stevepeterson, Stevouk, Stickee, Stikshift, StoneColdCrazy, StoneProphet, Stonnman, Stor stark7, Stormscape, Str1977, Stratd00d, StrongProudGunny, StuartBrady, Stuartrocks@gmail.com, Student7, Stuffit, Subseven26, Subtropical-man, Success123$, Suisui, Sullengrrl, SunDog, Sundar, Sundar1, Sunray, Supadawg, SuperHamster, Supertask, Supreme Deliciousness, Surlyduff50, Suruena, Suryoye, Sushi Leone, Sux2ussj, SvenskaJohannes, Swatjester, Swedish fusilier, Swid, Synthe, Syrthiss, Syslord2k, Szajci, Sümpf, T2Green, TFCforever, TGC55, THEunique, THX-1138, TSO1D, TShilo12, TUF-KAT, Ta bu shi da yu, Tacen28, Talorrob, Tangotango, Tapir Terrific, Targeman, Tariqabjotu, Tarret, Tartessos75, Tasc, TaurusMount, Taw, Tawker, Taxman, Tb, Tbacker, Tbhotch, Tcooke, Teccen, Tecomaster345, TedE, Teddycom, Tedp, Teh tennisman, Template namespace initialisation script, TenOfAllTrades, Terence, Teryx, Teuton9, Teutsch, TexasAndroid, Texture, Tezero, Tfine80, Tfjt, Thabet202, Thaejas, Thayts, The Anome, The Catfish, The Clinkynator, The Emirr, The Mysterious El Willstro, The Phoenix, The Thing That Should Not Be, The Tom, The Transhumanist, The Utahraptor, The demiurge, The dog2, The misha, The monkeyhate, The russian leader, The undertow, TheFox2, TheK, TheKMan, TheLewisRepublic, ThePointblank, TheTrojanHought, Theentireending, Thefullback, Thegoodson, Themanwithoutapast, Theonlyedge, Therealdrag0, Therequiembellishere, Thewallowmaker, Thewikipedian, Thing that goes on feet, Thingg, Think outside the box, Thomas Ganter, Thomas Ludwig, ThomasK, Thore, Threepwood89, Thricecube, Thughes1774, Thumperward, Thunk, Thuresson, Tiddle 999, Tiddly Tom, Tiki99, Tillwe, Tim buckley, TimBRoy, TimBentley, Timeineurope, Timmothias, Timwi, Tintenfischlein, Titoxd, Tiwonk, Tjl1128, Tkgd2007, Tkinias, Tkynerd, Tmasuda, Tobby72, Tobyw87, Tocino, Toffel07, Tokikake, Tokiohotelover, Tolaagboola, Tom harrison, TomR, TomaK, Tomeasy, Tomfreelance, Tomi, Tommy2010, Tommycw1, TomoK12, Tomorrow Never Knows, Tompayne1976, Tone, Tony Sidaway, Toothpull, Toussaint, TownDown, Tpbradbury, Tra, Trasman, Travelbird, Traveletti, TrbleClef, TreXiaNos, Tree Biting Conspiracy, Treisijs, Trewin, Treyt021, Trialsanderrors, Trickstar, Triggerpull2, Trilobite, Trinitrix, Troy 07, Trueshow111, Tsujigiri, Ttownfeen, Tulkolahten, Tunheim, Tuomas, TurabianNights, Twilight Princess 64, Twoshirou, Tyduro, Tyrol5, Tzartzam, Tzustrategy, U.S.A.U.S.A.U.S.A., U.Steele, UKWiki, UNDGRND, Uannis, Ubermonkey, Ubiquinoid, Ufwuct, Ugur Basak, UkPaolo, Ulf Heinsohn, Ulitzde, Ulmanor, Ulritz, Ultimate Destiny, Umbugbene, UncleAdolf, Underlying lk, Unify, Universe=atom, Unreal595, Unused0029, Unyoyega, Urban XII, Uriber, Ursutraide, User F203, User27091, UserDoe, Usergreatpower, Utcursch, Ute in DC, Uurtamo, V6g3h7, Valentinian, Valentino, Valleyhome, Van helsing, Vancouverguy, Vanderdecken, Vaniba12, Vanis, Vardion, Varlaam, Vary, Vdjj1960, Veit, Veledan, Velho, Ventur, Verdadero, Verman1, VeryVerily, Vexorg, Vextration, Victor Hugo, Viewfinder, Vilallonga, Vincentnufcr1, Violetriga, Vir, Virgule82, VitaleBaby, Vkem, VolatileChemical, Volunteer Marek, Vondort, Vortexx, Vranak, Vrghs jacob, Vsmith, Vulcant13, Vzbs34, WHEELER, WTGDMan1986, WTucker, Wackojacko120, Wafulz, Walkerma, Wangi, Waprap, Warfwar3, Warhead66, Warofdreams, Watbjbjbjk, Wayward, We hope, Weareqpr, WegianWarrior, Weissmann, Welle, Welshleprechaun, Wer900, WereWolf, Wernergerman, Westb3ach, Wester, Whaatt, WhisperToMe, Whitey115, Who, Whouk, Wiatr, Wik, Wiki alf, WikiBrown, Wikiain, Wikiborg, Wikifriendawesome, Wikipedia-WillYouMarryMe?, Wikipelli, Wikiperuvian, Wikist, Wikithor, Will Beback Auto, Willbur, Williamborg, Williame3, Williamwade, Willyhunter123, Wimt, Windsamurai, Winhunter, Wireless friend, WishingForever, Wissahickon Creek, Wk muriithi, Wknight94, Wliiam Shakespeare, Wlmg, Wolf530, Woohookitty, Wootini, Worm That Turned, Wow rules the world, WriterHound, WritersCramp, Wulf Isebrand, XLerate, XZise, Xaosflux, Xenus, Xeph, Xezbeth, Xigan, Xtopher, YAG490, Yahel Guhan, Yamamoto Ichiro, Yandman, YankeeDoodle14, YellowMonkey, Yngvadottir, Yoda1893, Yoghurt, Yoki11, Yorkshirian, Youknowyouknow, Youmuppet, Youngamerican, YourEyesOnly, Yousou, Youssefsan, Yvesdesprit, Yzmo, Z.E.R.O., Zacefron52, Zack Holly Venturi, Zahnradzacken, Zaindy87, ZakuSage, Zaphodia, Zarbi1, Zarniwoot, Zeldafan3242, Zeno Gantner, Zesel, Zia demion, Ziusudra, Zmaj, Zondor, Zoney, Zpatel93, Zscout370, Zsinj, Zundark, Zzuuzz, Zzyzx11, £20, °, Äpple, Ævar Arnfjörð Bjarmason, Île flottante, Özdemir, ΚΕΚΡΩΨ, Методије, ירעל ירניהארט, اپسیلون, رسن ربلین, €pa, , 4458 anonymous edits

Image Sources, Licenses and Contributors

file:Germany location map.svg *Source*: http://en.wikipedia.org/w/index.php?title=File:Germany_location_map.svg *License*: unknown *Contributors*: User:NordNordWest

File:Red_pog.svg *Source*: http://en.wikipedia.org/w/index.php?title=File:Red_pog.svg *License*: unknown *Contributors*: Anomie

Image:Administrative Gliederung Deutschlands.svg *Source*: http://en.wikipedia.org/w/index.php?title=File:Administrative_Gliederung_Deutschlands.svg *License*: unknown *Contributors*: User:glglgl

Image:WeilheimChurch.JPG *Source*: http://en.wikipedia.org/w/index.php?title=File:WeilheimChurch.JPG *License*: unknown *Contributors*: User:SreeBot

File:Wappen Weilheim an der Teck.png *Source*: http://en.wikipedia.org/w/index.php?title=File:Wappen_Weilheim_an_der_Teck.png *License*: unknown *Contributors*: User Augapfel on de.wikipedia

File:Administrative Districts of Germany 1981-2008.svg *Source*: http://en.wikipedia.org/w/index.php?title=File:Administrative_Districts_of_Germany_1981-2008.svg *License*: unknown *Contributors*: Bennet Schulte

File:Flag of Baden-Württemberg.svg *Source*: http://en.wikipedia.org/w/index.php?title=File:Flag_of_Baden-Württemberg.svg *License*: unknown *Contributors*: Burts, Cycn, Dbenbenn, Fry1989, Kookaburra, Ludger1961, Madden, Mattes, Mogelzahn, Ms2ger, Phlegmatic, Pilettes, Rtc, Sarang, Ssch

Image:Wirkoennenalles.jpg *Source*: http://en.wikipedia.org/w/index.php?title=File:Wirkoennenalles.jpg *License*: unknown *Contributors*: Scholz & Friends

File:StuttgartSchlossPlatz.JPG *Source*: http://en.wikipedia.org/w/index.php?title=File:StuttgartSchlossPlatz.JPG *License*: unknown *Contributors*: User:BuzzWoof

File:Karlsruhe Germany CastleByNight.jpg *Source*: http://en.wikipedia.org/w/index.php?title=File:Karlsruhe_Germany_CastleByNight.jpg *License*: unknown *Contributors*: Ikar.us, Olivier2, Rüdiger Wölk, Sitacuisses, Wikipedia-ce, 2 anonymous edits

File:Heidelberg 20060420 021.jpg *Source*: http://en.wikipedia.org/w/index.php?title=File:Heidelberg_20060420_021.jpg *License*: unknown *Contributors*: Frank-m, Hlohning, NeverDoING

File:Freiburg view.jpg *Source*: http://en.wikipedia.org/w/index.php?title=File:Freiburg_view.jpg *License*: unknown *Contributors*: Peter Andersen

File:Universitaet Mannheim Schloss Ehrenhof.jpg *Source*: http://en.wikipedia.org/w/index.php?title=File:Universitaet_Mannheim_Schloss_Ehrenhof.jpg *License*: unknown *Contributors*: User:Stefanie Eichler

File:Ulm - Stadtkulisse Ulm mit Donau.jpg *Source*: http://en.wikipedia.org/w/index.php?title=File:Ulm_-_Stadtkulisse_Ulm_mit_Donau.jpg *License*: unknown *Contributors*: User:Franzfoto

File:View of Baden-baden.jpeg *Source*: http://en.wikipedia.org/w/index.php?title=File:View_of_Baden-baden.jpeg *License*: unknown *Contributors*: User:Kirillll

File:Pforzheim Wallberg.jpg *Source*: http://en.wikipedia.org/w/index.php?title=File:Pforzheim_Wallberg.jpg *License*: unknown *Contributors*: Jacques Grießmayer

File:Cities and Districts in Baden-Wuerttemberg.svg *Source*: http://en.wikipedia.org/w/index.php?title=File:Cities_and_Districts_in_Baden-Wuerttemberg.svg *License*: unknown *Contributors*: User:AxG

File:Uni KA Portal.jpg *Source*: http://en.wikipedia.org/w/index.php?title=File:Uni_KA_Portal.jpg *License*: unknown *Contributors*: Original uploader was AlterVista at de.wikipedia. Later version(s) were uploaded by Sitacuisses at de.wikipedia.

File:Openstreetmap logo.svg *Source*: http://en.wikipedia.org/w/index.php?title=File:Openstreetmap_logo.svg *License*: unknown *Contributors*: OpenStreetMap

File:Germany general map.png *Source*: http://en.wikipedia.org/w/index.php?title=File:Germany_general_map.png *License*: unknown *Contributors*: User:Lencer, User:NordNordWest

File:Flag of Belgium (civil).svg *Source*: http://en.wikipedia.org/w/index.php?title=File:Flag_of_Belgium_(civil).svg *License*: unknown *Contributors*: Bean49, David Descamps, Dbenbenn, Denelson83, Evanc0912, Fry1989, Gabriel trzy, Howcome, IvanOS, Ms2ger, Nightstallion, Oreo Priest, Rocket000, Rodejong, Sir Iain, ThomasPusch, Warddr, Zscout370, 4 anonymous edits

File:Flag of France.svg *Source*: http://en.wikipedia.org/w/index.php?title=File:Flag_of_France.svg *License*: unknown *Contributors*: Anomie

File:Flag of the United Kingdom.svg *Source*: http://en.wikipedia.org/w/index.php?title=File:Flag_of_the_United_Kingdom.svg *License*: unknown *Contributors*: Anomie, Good Olfactory, Mifter

File:Flag of Spain.svg *Source*: http://en.wikipedia.org/w/index.php?title=File:Flag_of_Spain.svg *License*: unknown *Contributors*: Anomie

File:Flag of the People's Republic of China.svg *Source*: http://en.wikipedia.org/w/index.php?title=File:Flag_of_the_People's_Republic_of_China.svg *License*: unknown *Contributors*: User:Denelson83, User:SKopp, User:Shizhao, User:Zscout370

File:Flag of Germany.svg *Source*: http://en.wikipedia.org/w/index.php?title=File:Flag_of_Germany.svg *License*: unknown *Contributors*: Anomie

File:Flag of the United States.svg *Source*: http://en.wikipedia.org/w/index.php?title=File:Flag_of_the_United_States.svg *License*: unknown *Contributors*: Anomie

File:Flag of South Africa.svg *Source*: http://en.wikipedia.org/w/index.php?title=File:Flag_of_South_Africa.svg *License*: unknown *Contributors*: Adriaan, Anime Addict AA, AnonMoos, BRUTE, Daemonic Kangaroo, Dnik, Duduziq, Dzordzm, Fry1989, Homo lupus, Jappalang, Juliancolton, Kam Solusar, Klemen Kocjancic, Klymene, Lexxyy, Mahahahaneapneap, Manuelt15, Moviedefender, NeverDoING, Ninane, Poznaniak, Przemub, SKopp, Sarang, SiBr4, ThePCKid, ThomasPusch, Tvdm, Ultratomio, Vzb83, Zscout370, 35 anonymous edits

File:Flag of Russia.svg *Source*: http://en.wikipedia.org/w/index.php?title=File:Flag_of_Russia.svg *License*: unknown *Contributors*: Anomie, Zscout370

File:Flag of Israel.svg *Source*: http://en.wikipedia.org/w/index.php?title=File:Flag_of_Israel.svg *License*: unknown *Contributors*: "The Provisional Council of State Proclamation of the Flag of the State of Israel" of 25 Tishrei 5709 (28 October 1948) provides the official specification for the design of the Israeli flag. The color of the Magen David and the stripes of the Israeli flag is not precisely specified by the above legislation. The color depicted in the current version of the image is typical of flags used in Israel today, although individual flags can and do vary. The flag legislation officially specifies dimensions of 220 cm × 160 cm. However, the sizes of actual flags vary (although the aspect ratio is usually retained).

File:Flag of Ireland.svg *Source*: http://en.wikipedia.org/w/index.php?title=File:Flag_of_Ireland.svg *License*: unknown *Contributors*: User:SKopp

File:Flag of Hungary.svg *Source*: http://en.wikipedia.org/w/index.php?title=File:Flag_of_Hungary.svg *License*: unknown *Contributors*: User:SKopp

File:Flag of Turkey.svg *Source*: http://en.wikipedia.org/w/index.php?title=File:Flag_of_Turkey.svg *License*: unknown *Contributors*: User:Dbenbenn

File:Flag of Italy.svg *Source*: http://en.wikipedia.org/w/index.php?title=File:Flag_of_Italy.svg *License*: unknown *Contributors*: Anomie

File:Flag of Greece.svg *Source*: http://en.wikipedia.org/w/index.php?title=File:Flag_of_Greece.svg *License*: unknown *Contributors*: (of code) (talk)

File:Flag of Malta.svg *Source*: http://en.wikipedia.org/w/index.php?title=File:Flag_of_Malta.svg *License*: unknown *Contributors*: Alkari, Cycn, Fry1989, Gabbe, Homo lupus, Klemen Kocjancic, Liftarn, Mattes, Meno25, Nightstallion, Peeperman, Pumbaa80, Ratatosk, Rodejong, Zscout370, 5 anonymous edits

File:Flag of Luxembourg.svg *Source*: http://en.wikipedia.org/w/index.php?title=File:Flag_of_Luxembourg.svg *License*: unknown *Contributors*: User:SKopp

File:Flag of Estonia.svg *Source*: http://en.wikipedia.org/w/index.php?title=File:Flag_of_Estonia.svg *License*: unknown *Contributors*: User:PeepP, User:SKopp

File:Flag of Austria.svg *Source*: http://en.wikipedia.org/w/index.php?title=File:Flag_of_Austria.svg *License*: unknown *Contributors*: User:SKopp

File:Flag of Denmark.svg *Source*: http://en.wikipedia.org/w/index.php?title=File:Flag_of_Denmark.svg *License*: unknown *Contributors*: User:Madden

File:Flag of Slovakia.svg *Source*: http://en.wikipedia.org/w/index.php?title=File:Flag_of_Slovakia.svg *License*: unknown *Contributors*: User:SKopp

File:Flag of Poland.svg *Source*: http://en.wikipedia.org/w/index.php?title=File:Flag_of_Poland.svg *License*: unknown *Contributors*: Anomie, Mifter

File:Flag of Slovenia.svg *Source*: http://en.wikipedia.org/w/index.php?title=File:Flag_of_Slovenia.svg *License*: unknown *Contributors*: User:Achim1999

File:Flag of Finland.svg *Source*: http://en.wikipedia.org/w/index.php?title=File:Flag_of_Finland.svg *License*: unknown *Contributors*: User:SKopp

File:Flag of the Czech Republic.svg *Source*: http://en.wikipedia.org/w/index.php?title=File:Flag_of_the_Czech_Republic.svg *License*: unknown *Contributors*: special commission (of code): SVG version by cs:-xfi-. Colors according to Appendix No. 3 of czech legal Act 3/1993. cs:Zirland.

File:Flag of Latvia.svg *Source*: http://en.wikipedia.org/w/index.php?title=File:Flag_of_Latvia.svg *License*: unknown *Contributors*: User:SKopp

File:Flag of Sweden.svg *Source*: http://en.wikipedia.org/w/index.php?title=File:Flag_of_Sweden.svg *License*: unknown *Contributors*: Anomie

File:Flag of the Netherlands.svg *Source*: http://en.wikipedia.org/w/index.php?title=File:Flag_of_the_Netherlands.svg *License*: unknown *Contributors*: User:Zscout370

File:Flag of Romania.svg *Source*: http://en.wikipedia.org/w/index.php?title=File:Flag_of_Romania.svg *License*: unknown *Contributors*: User:AdiJapan

File:Flag of Colombia.svg *Source*: http://en.wikipedia.org/w/index.php?title=File:Flag_of_Colombia.svg *License*: unknown *Contributors*: User:SKopp

File:Flag of Scotland.svg *Source*: http://en.wikipedia.org/w/index.php?title=File:Flag_of_Scotland.svg *License*: unknown *Contributors*: User:Kbolino

File:Flag of Japan.svg *Source*: http://en.wikipedia.org/w/index.php?title=File:Flag_of_Japan.svg *License*: unknown *Contributors*: Anomie

File:Flag of Algeria.svg *Source*: http://en.wikipedia.org/w/index.php?title=File:Flag_of_Algeria.svg *License*: unknown *Contributors*: User:SKopp

File:Flag of Switzerland.svg *Source*: http://en.wikipedia.org/w/index.php?title=File:Flag_of_Switzerland.svg *License*: unknown *Contributors*: User:-xfi-, User:Marc Mongenet, User:Zscout370

File:Flag of Croatia.svg *Source*: http://en.wikipedia.org/w/index.php?title=File:Flag_of_Croatia.svg *License*: unknown *Contributors*: User:MaGa

File:Flag of Serbia.svg *Source*: http://en.wikipedia.org/w/index.php?title=File:Flag_of_Serbia.svg *License*: unknown *Contributors*: sodipodi.com

File:Flag of Portugal.svg *Source*: http://en.wikipedia.org/w/index.php?title=File:Flag_of_Portugal.svg *License*: unknown *Contributors*: Columbano Bordalo Pinheiro (1910; generic design); Vítor Luís Rodrigues; António Martins-Tuválkin (2004; this specific vector set: see sources)

File:Flag of Lithuania.svg *Source*: http://en.wikipedia.org/w/index.php?title=File:Flag_of_Lithuania.svg *License*: unknown *Contributors*: User:SKopp

File:Flag of Cyprus.svg *Source*: http://en.wikipedia.org/w/index.php?title=File:Flag_of_Cyprus.svg *License*: unknown *Contributors*: User:Vzb83

File:Flag of Bulgaria.svg *Source*: http://en.wikipedia.org/w/index.php?title=File:Flag_of_Bulgaria.svg *License*: unknown *Contributors*: User:SKopp

File:Flag of Wales 2.svg *Source*: http://en.wikipedia.org/w/index.php?title=File:Flag_of_Wales_2.svg *License*: unknown *Contributors*: AlexD, Bjankuloski06en, Cecil, Cycn, Dbenbenn, Duduziq, F. F. Fjodor, FruitMonkey, Fry1989, Homo lupus, Iago4096, IvanOS, Pumbaa80, Red devil 666, Srtxg, Tha real, Torstein, Vernanimalcula, Vzb83, Wouterhagens, 7 anonymous edits

File:Flag of Kenya.svg *Source*: http://en.wikipedia.org/w/index.php?title=File:Flag_of_Kenya.svg *License*: unknown *Contributors*: User:Pumbaa80

File:Flag of England.svg *Source*: http://en.wikipedia.org/w/index.php?title=File:Flag_of_England.svg *License*: unknown *Contributors*: Anomie

File:Flag of Ukraine.svg *Source*: http://en.wikipedia.org/w/index.php?title=File:Flag_of_Ukraine.svg *License*: unknown *Contributors*: User:Jon Harald Søby, User:Zscout370

File:Flag of Palestine.svg *Source*: http://en.wikipedia.org/w/index.php?title=File:Flag_of_Palestine.svg *License*: unknown *Contributors*: User:Orionist, user:Makaristos

File:Flag of Indonesia.svg *Source*: http://en.wikipedia.org/w/index.php?title=File:Flag_of_Indonesia.svg *License*: unknown *Contributors*: User:Gabbe, User:SKopp

File:Flag of Uzbekistan.svg *Source*: http://en.wikipedia.org/w/index.php?title=File:Flag_of_Uzbekistan.svg *License*: unknown *Contributors*: User:Zscout370

File:Flag of Mexico.svg *Source*: http://en.wikipedia.org/w/index.php?title=File:Flag_of_Mexico.svg *License*: unknown *Contributors*: User:AlexCovarrubias

File:Flag of Argentina.svg *Source*: http://en.wikipedia.org/w/index.php?title=File:Flag_of_Argentina.svg *License*: unknown *Contributors*: User:Dbenbenn

File:Flag of Brazil.svg *Source*: http://en.wikipedia.org/w/index.php?title=File:Flag_of_Brazil.svg *License*: unknown *Contributors*: Anomie

File:Flag of Namibia.svg *Source*: http://en.wikipedia.org/w/index.php?title=File:Flag_of_Namibia.svg *License*: unknown *Contributors*: User:Vzb83

File:Flag of Iran.svg *Source*: http://en.wikipedia.org/w/index.php?title=File:Flag_of_Iran.svg *License*: unknown *Contributors*: Various

File:Flag of Nicaragua.svg *Source*: http://en.wikipedia.org/w/index.php?title=File:Flag_of_Nicaragua.svg *License*: unknown *Contributors*: User:Nightstallion

File:Flag of Mozambique.svg *Source*: http://en.wikipedia.org/w/index.php?title=File:Flag_of_Mozambique.svg *License*: unknown *Contributors*: User:Nightstallion

File:Flag of Belarus.svg *Source*: http://en.wikipedia.org/w/index.php?title=File:Flag_of_Belarus.svg *License*: unknown *Contributors*: User:Zscout370

File:Flag of Norway.svg *Source*: http://en.wikipedia.org/w/index.php?title=File:Flag_of_Norway.svg *License*: unknown *Contributors*: User:Dbenbenn

File:Flag of Burkina Faso.svg *Source*: http://en.wikipedia.org/w/index.php?title=File:Flag_of_Burkina_Faso.svg *License*: unknown *Contributors*: User:Gabbe, User:SKopp

File:Flag of Peru.svg *Source*: http://en.wikipedia.org/w/index.php?title=File:Flag_of_Peru.svg *License*: unknown *Contributors*: User:Dbenbenn

File:Flag of Georgia.svg *Source*: http://en.wikipedia.org/w/index.php?title=File:Flag_of_Georgia.svg *License*: unknown *Contributors*: User:SKopp

File:Flag of Bolivia.svg *Source*: http://en.wikipedia.org/w/index.php?title=File:Flag_of_Bolivia.svg *License*: unknown *Contributors*: User:SKopp

File:Flag of Ghana.svg *Source*: http://en.wikipedia.org/w/index.php?title=File:Flag_of_Ghana.svg *License*: unknown *Contributors*: Benchill, Cycn, Fry1989, Henswick, Homo lupus, Indolences, Jarekt, Klemen Kocjancic, Magasjukur2, Neq00, OAlexander, SKopp, ThomasPusch, Threecharlie, Torstein, Zscout370, 5 anonymous edits

File:Flag of Mongolia.svg *Source*: http://en.wikipedia.org/w/index.php?title=File:Flag_of_Mongolia.svg *License*: unknown *Contributors*: User:Zscout370

File:Flag of Tunisia.svg *Source*: http://en.wikipedia.org/w/index.php?title=File:Flag_of_Tunisia.svg *License*: unknown *Contributors*: entraîneur: BEN KHALIFA WISSAM

File:Flag of Mali.svg *Source*: http://en.wikipedia.org/w/index.php?title=File:Flag_of_Mali.svg *License*: unknown *Contributors*: User:SKopp

File:Flag of Canada.svg *Source*: http://en.wikipedia.org/w/index.php?title=File:Flag_of_Canada.svg *License*: unknown *Contributors*: Anomie

File:Flag of Greenland.svg *Source*: http://en.wikipedia.org/w/index.php?title=File:Flag_of_Greenland.svg *License*: unknown *Contributors*: Jeffrey Connell (IceKarma)

File:Flag of Iceland.svg *Source*: http://en.wikipedia.org/w/index.php?title=File:Flag_of_Iceland.svg *License*: unknown *Contributors*: User:Zscout370, User:Ævar Arnfjörð Bjarmason

File:Flag of Paraguay.svg *Source*: http://en.wikipedia.org/w/index.php?title=File:Flag_of_Paraguay.svg *License*: unknown *Contributors*: Republica del Paraguay

File:Flag of Macedonia.svg *Source*: http://en.wikipedia.org/w/index.php?title=File:Flag_of_Macedonia.svg *License*: unknown *Contributors*: User:Gabbe, User:SKopp

File:Flag of the Republic of the Congo.svg *Source*: http://en.wikipedia.org/w/index.php?title=File:Flag_of_the_Republic_of_the_Congo.svg *License*: unknown *Contributors*: Anime Addict AA, Antemister, Courcelles, Denelson83, Erlenmeyer, Estrilda, FischersFritz, Fry1989, Homo lupus, Klemen Kocjancic, LA2, Madden, Mattes, Moyogo, Neq00, Nightstallion, Persiana, Ratatosk, Romaine, ThomasPusch, Thuresson, 3 anonymous edits

File:Flag of Togo.svg *Source*: http://en.wikipedia.org/w/index.php?title=File:Flag_of_Togo.svg *License*: unknown *Contributors*: Aaker, Ahsoous, Alkari, Cycn, EugeneZelenko, Fry1989, Homo lupus, Klemen Kocjancic, Mattes, Mxn, Neq00, Nightstallion, Reisio, ThomasPusch, Vzb83

File:Flag of Honduras.svg *Source*: http://en.wikipedia.org/w/index.php?title=File:Flag_of_Honduras.svg *License*: unknown *Contributors*: D1990, Denelson83, ECanalla, Feydey, Fred J, Homo lupus, Klemen Kocjancic, Mattes, Matthew hk, Neq00, Oak27, Pumbaa80, Rocket000, RubiksMaster110, SKopp, ThomasPusch, Tocino, Vzb83, Yuval Madar, ZooFari, Zscout370, 10 anonymous edits

File:Flag of Bosnia and Herzegovina.svg *Source*: http://en.wikipedia.org/w/index.php?title=File:Flag_of_Bosnia_and_Herzegovina.svg *License*: unknown *Contributors*: User:Kseferovic

File:Flag of Tanzania.svg *Source*: http://en.wikipedia.org/w/index.php?title=File:Flag_of_Tanzania.svg *License*: unknown *Contributors*: User:Alkari, User:Madden, User:SKopp

File:Flag of Moldova.svg *Source*: http://en.wikipedia.org/w/index.php?title=File:Flag_of_Moldova.svg *License*: unknown *Contributors*: User:Nameneko

File:Flag of Venezuela.svg *Source*: http://en.wikipedia.org/w/index.php?title=File:Flag_of_Venezuela.svg *License*: unknown *Contributors*: Alkari, Bastique, Cycn, Denelson83, DerFussi, Fry1989, George McFinnigan, Herbythyme, Homo lupus, Huhsunqu, Infrogmation, K21edgo, Klemen Kocjancic, Ludger1961, Neq00, Nightstallion, Reisio, Rupert Pupkin, Sarang, Sparkve, ThomasPusch, Vzb83, Wikisole, Zscout370, 12 anonymous edits

File:Flag of India.svg *Source*: http://en.wikipedia.org/w/index.php?title=File:Flag_of_India.svg *License*: unknown *Contributors*: Anomie, Mifter

File:Flag of Malawi.svg *Source*: http://en.wikipedia.org/w/index.php?title=File:Flag_of_Malawi.svg *License*: unknown *Contributors*: Achim1999, AnonMoos, Antonsusi, Awadewit, Erlenmeyer, Fred J, Fry1989, Gddea, Homo lupus, IvanLanin, Klemen Kocjancic, Mattes, Phlegmatic, Rodejong, SKopp, Sarang, Sebjarod, Sweeper tamonten, ThomasPusch, Zscout370, 4 anonymous edits

File:Flag of Egypt.svg *Source*: http://en.wikipedia.org/w/index.php?title=File:Flag_of_Egypt.svg *License*: unknown *Contributors*: Open Clip Art

File:Flag of Albania.svg *Source*: http://en.wikipedia.org/w/index.php?title=File:Flag_of_Albania.svg *License*: unknown *Contributors*: User:Dbenbenn

File:Flag of Montenegro.svg *Source*: http://en.wikipedia.org/w/index.php?title=File:Flag_of_Montenegro.svg *License*: unknown *Contributors*: User:B1mbo, User:Froztbyte

File:Flag of Costa Rica.svg *Source*: http://en.wikipedia.org/w/index.php?title=File:Flag_of_Costa_Rica.svg *License*: unknown *Contributors*: User:Gabbe, User:SKopp

File:Flag of Rwanda.svg *Source*: http://en.wikipedia.org/w/index.php?title=File:Flag_of_Rwanda.svg *License*: unknown *Contributors*: .

File:Flag of Ethiopia.svg *Source*: http://en.wikipedia.org/w/index.php?title=File:Flag_of_Ethiopia.svg *License*: unknown *Contributors*: Aaker, Anime Addict AA, Antemister, Cycn, F l a n k e r, Fry1989, GoodMorningEthiopia, Happenstance, Homo lupus, Huhsunqu, Ixfd64, Klemen Kocjancic, MartinThoma, Mattes, Mozzan, Neq00, OAlexander, Pumbaa80, Rainforest tropicana, Reisio, SKopp, Smooth O, Spiritia, ThomasPusch, Torstein, Wsiegmund, Zscout370, 16 anonymous edits

File:Flag of Azerbaijan.svg *Source*: http://en.wikipedia.org/w/index.php?title=File:Flag_of_Azerbaijan.svg *License*: unknown *Contributors*: User:SKopp

File:Flag of Zimbabwe.svg *Source*: http://en.wikipedia.org/w/index.php?title=File:Flag_of_Zimbabwe.svg *License*: unknown *Contributors*: User:Madden

File:Flag of Australia.svg *Source*: http://en.wikipedia.org/w/index.php?title=File:Flag_of_Australia.svg *License*: unknown *Contributors*: Anomie, Mifter

File:Flag of South Korea.svg *Source*: http://en.wikipedia.org/w/index.php?title=File:Flag_of_South_Korea.svg *License*: unknown *Contributors*: Various

File:Flag of Cameroon.svg *Source*: http://en.wikipedia.org/w/index.php?title=File:Flag_of_Cameroon.svg *License*: unknown *Contributors*: (of code)

File:Flag of the Philippines.svg *Source*: http://en.wikipedia.org/w/index.php?title=File:Flag_of_the_Philippines.svg *License*: unknown *Contributors*: User:Achim1999

File:Flag of Bavaria (lozengy).svg *Source*: http://en.wikipedia.org/w/index.php?title=File:Flag_of_Bavaria_(lozengy).svg *License*: unknown *Contributors*: diese Datei: Jwnabd

File:Coat of arms of Bavaria.svg *Source*: http://en.wikipedia.org/w/index.php?title=File:Coat_of_arms_of_Bavaria.svg *License*: unknown *Contributors*: Freistaat Bayern

File:Deutschland Lage von Bayern.svg *Source*: http://en.wikipedia.org/w/index.php?title=File:Deutschland_Lage_von_Bayern.svg *License*: unknown *Contributors*: User:David Liuzzo

File:Speaker Icon.svg *Source*: http://en.wikipedia.org/w/index.php?title=File:Speaker_Icon.svg *License*: unknown *Contributors*: Blast, G.Hagedorn, Mobius, Tehdog, 2 anonymous edits

File:Heunischenburg.jpg *Source*: http://en.wikipedia.org/w/index.php?title=File:Heunischenburg.jpg *License*: unknown *Contributors*: Stefan Wicklein, Kronach

File:Bayern im 10.Jh.png *Source*: http://en.wikipedia.org/w/index.php?title=File:Bayern_im_10.Jh.png *License*: unknown *Contributors*: Electionworld, Ephraim33, Koryakov Yuri, Rfortner, 1 anonymous edits

File:Bayern nach der Teilung 1392.png *Source*: http://en.wikipedia.org/w/index.php?title=File:Bayern_nach_der_Teilung_1392.png *License*: unknown *Contributors*: User:Lencer

File:Bayern von 1800 bis heute.png *Source*: http://en.wikipedia.org/w/index.php?title=File:Bayern_von_1800_bis_heute.png *License*: unknown *Contributors*: Original uploader was Tk at de.wikipedia

File:Block of Bavarian stamps (1920s) overprinted with "Deutsches Reich".jpg *Source*: http://en.wikipedia.org/w/index.php?title=File:Block_of_Bavarian_stamps_(1920s)_overprinted_with_"Deutsches_Reich".jpg *License*: unknown *Contributors*: User:Atzmonit

File:Kriegerdenkmal Dietelskirchen.jpg *Source*: http://en.wikipedia.org/w/index.php?title=File:Kriegerdenkmal_Dietelskirchen.jpg *License*: unknown *Contributors*: Seppal

File:Bavarian Herald.jpg *Source*: http://en.wikipedia.org/w/index.php?title=File:Bavarian_Herald.jpg *License*: unknown *Contributors*: Evadb, FordPrefect42, Mattes, Mogelzahn, 2 anonymous edits

File:Bavarian Alps 2002.jpg *Source*: http://en.wikipedia.org/w/index.php?title=File:Bavarian_Alps_2002.jpg *License*: unknown *Contributors*: Melancholia i

File:Bavarian Admin Districts.jpg *Source*: http://en.wikipedia.org/w/index.php?title=File:Bavarian_Admin_Districts.jpg *License*: unknown *Contributors*: Pere prlpz, TUBS

File:Karte-Bayern-Landkreise.png *Source*: http://en.wikipedia.org/w/index.php?title=File:Karte-Bayern-Landkreise.png *License*: unknown *Contributors*: User Cordyph on en.wikipedia

File:Rathaus and Marienplatz from Peterskirche - August 2006.jpg *Source*: http://en.wikipedia.org/w/index.php?title=File:Rathaus_and_Marienplatz_from_Peterskirche_-_August_2006.jpg *License*: unknown *Contributors*: User:Diliff

File:Nürnberg panorama.jpg *Source*: http://en.wikipedia.org/w/index.php?title=File:Nürnberg_panorama.jpg *License*: unknown *Contributors*: 34er, AndreasPraefcke, BlackIceNRW, Demidow, Mb1302, Technokrat

File:Augsburger Rathausplatz.jpg *Source*: http://en.wikipedia.org/w/index.php?title=File:Augsburger_Rathausplatz.jpg *License*: unknown *Contributors*: Vitold Muratov

File:Stadtansicht Regensburg.JPG *Source*: http://en.wikipedia.org/w/index.php?title=File:Stadtansicht_Regensburg.JPG *License*: unknown *Contributors*: HH58

File:Würzburg view.jpg *Source*: http://en.wikipedia.org/w/index.php?title=File:Würzburg_view.jpg *License*: unknown *Contributors*: Burny, FlickrLickr, Ronaldino, Till.niermann, Torben

File:Marktplatz Bayreuth.JPG *Source*: http://en.wikipedia.org/w/index.php?title=File:Marktplatz_Bayreuth.JPG *License*: unknown *Contributors*: User:Michael Sander

File:Seehofer.JPG *Source*: http://en.wikipedia.org/w/index.php?title=File:Seehofer.JPG *License*: unknown *Contributors*: User:J. Patrick Fischer

File:Germany bavaria alps-church.jpg *Source*: http://en.wikipedia.org/w/index.php?title=File:Germany_bavaria_alps-church.jpg *License*: unknown *Contributors*: User:RRKennison

File:Karwendel-Ahornboden.jpg *Source*: http://en.wikipedia.org/w/index.php?title=File:Karwendel-Ahornboden.jpg *License*: unknown *Contributors*: Herzi Pinki, Shyam, Svíčková, Zinnmann

File:Upper German.png *Source*: http://en.wikipedia.org/w/index.php?title=File:Upper_German.png *License*: unknown *Contributors*: User:Brichtig

File:Allianz arena at night Richard Bartz.jpg *Source*: http://en.wikipedia.org/w/index.php?title=File:Allianz_arena_at_night_Richard_Bartz.jpg *License*: unknown *Contributors*: User:Makro Freak, User:Richard Bartz

File:Aschaffenburg Schloss Johannisburg.jpg *Source*: http://en.wikipedia.org/w/index.php?title=File:Aschaffenburg_Schloss_Johannisburg.jpg *License*: unknown *Contributors*: User:Steschke

File:Wuerzburger Residenz vom Hofgarten.jpg *Source*: http://en.wikipedia.org/w/index.php?title=File:Wuerzburger_Residenz_vom_Hofgarten.jpg *License*: unknown *Contributors*: Wolfram Esser

File:Marienberg wuerzburg.jpg *Source*: http://en.wikipedia.org/w/index.php?title=File:Marienberg_wuerzburg.jpg *License*: unknown *Contributors*: Christian Horvat

File:Plassenburg oben.jpg *Source*: http://en.wikipedia.org/w/index.php?title=File:Plassenburg_oben.jpg *License*: unknown *Contributors*: Diabas, Moros, Schubbay, Sir Gawain

File:BambergDom.jpg *Source*: http://en.wikipedia.org/w/index.php?title=File:BambergDom.jpg *License*: unknown *Contributors*: AnRo0002, Kjetil r, Kurpfalzbilder.de, Maksim, Para, Schubbay, 1 anonymous edits

File:Vierzehnheiligen I.JPG *Source*: http://en.wikipedia.org/w/index.php?title=File:Vierzehnheiligen_I.JPG *License*: unknown *Contributors*: Asio otus, Ronaldino, Schubbay, Yarl

File:Coburg-Veste1.jpg *Source*: http://en.wikipedia.org/w/index.php?title=File:Coburg-Veste1.jpg *License*: unknown *Contributors*: User:Störfix

File:Bayreuth Festspielhaus 2006-07-16.jpg *Source*: http://en.wikipedia.org/w/index.php?title=File:Bayreuth_Festspielhaus_2006-07-16.jpg *License*: unknown *Contributors*: Rico Neitzel

File:Nuremberg sebald castle f lorenz f s.jpg *Source*: http://en.wikipedia.org/w/index.php?title=File:Nuremberg_sebald_castle_f_lorenz_f_s.jpg *License*: unknown *Contributors*: User:Keichwa

File:Kastell Biriciana (Weißenburg in Bayern).jpg *Source*: http://en.wikipedia.org/w/index.php?title=File:Kastell_Biriciana_(Weißenburg_in_Bayern).jpg *License*: unknown *Contributors*: Bapti, Brego, Breugelius, Cherubino, Clio20, Cristiano64, Umherirrender, Xenophon

File:Kreuztor Ingoldstadt.jpg *Source*: http://en.wikipedia.org/w/index.php?title=File:Kreuztor_Ingoldstadt.jpg *License*: unknown *Contributors*: AnRo0002, Androl, Cresi, Franconia, Fransvannes, Haui, High Contrast

File:Schloss Neuburg.jpg *Source*: http://en.wikipedia.org/w/index.php?title=File:Schloss_Neuburg.jpg *License*: unknown *Contributors*: Original uploader was Mendli at de.wikipedia

File:Regensburg-steinerne-bruecke-hytrion-enhanced_1-1024x768.jpg *Source*: http://en.wikipedia.org/w/index.php?title=File:Regensburg-steinerne-bruecke-hytrion-enhanced_1-1024x768.jpg *License*: unknown *Contributors*: original by revised by Wolpertinger on de.wikipedia

File:Walhalla aussen.jpg *Source*: http://en.wikipedia.org/w/index.php?title=File:Walhalla_aussen.jpg *License*: unknown *Contributors*: Tebdi

File:Befreiungshalle-kelheim-aussen.jpg *Source*: http://en.wikipedia.org/w/index.php?title=File:Befreiungshalle-kelheim-aussen.jpg *License*: unknown *Contributors*: Common Good, Magog the Ogre, NordNordWest

File:Passau Inn Cathedral.JPG *Source*: http://en.wikipedia.org/w/index.php?title=File:Passau_Inn_Cathedral.JPG *License*: unknown *Contributors*: Aconcagua, Priwo, Snommoc

File:LandshutTrausnitz01.jpg *Source*: http://en.wikipedia.org/w/index.php?title=File:LandshutTrausnitz01.jpg *License*: unknown *Contributors*: Alexander Z., Rup11, Sir Gawain

File:Burghausen.jpg *Source*: http://en.wikipedia.org/w/index.php?title=File:Burghausen.jpg *License*: unknown *Contributors*: Reinhard Kraasch

File:A rathausplatz.jpg *Source*: http://en.wikipedia.org/w/index.php?title=File:A_rathausplatz.jpg *License*: unknown *Contributors*: Sven Jansen. User Spielkind on de.wikipedia

File:Munich skyline.jpg *Source*: http://en.wikipedia.org/w/index.php?title=File:Munich_skyline.jpg *License*: unknown *Contributors*: User:Stefan Kühn

File:Residenz Ansicht Hofgarten, München.jpg *Source*: http://en.wikipedia.org/w/index.php?title=File:Residenz_Ansicht_Hofgarten,_München.jpg *License*: unknown *Contributors*: User:Gryffindor

File:Image-Schloss Nymphenburg Munich CC edit3.jpg *Source*: http://en.wikipedia.org/w/index.php?title=File:Image-Schloss_Nymphenburg_Munich_CC_edit3.jpg *License*: unknown *Contributors*: User:Makro Freak, User:Richard Bartz

File:Freisinger Dom aussen 01.jpg *Source*: http://en.wikipedia.org/w/index.php?title=File:Freisinger_Dom_aussen_01.jpg *License*: unknown *Contributors*: Lmmrs

File:Herrenchiem.JPG *Source*: http://en.wikipedia.org/w/index.php?title=File:Herrenchiem.JPG *License*: unknown *Contributors*: Mattis, Sir Gawain, Würfel, 1 anonymous edits

File:Linderhof-1.jpg *Source*: http://en.wikipedia.org/w/index.php?title=File:Linderhof-1.jpg *License*: unknown *Contributors*: User:Softeis

File:Schloss Hohenschwangau.jpg *Source*: http://en.wikipedia.org/w/index.php?title=File:Schloss_Hohenschwangau.jpg *License*: unknown *Contributors*: user:Transpassive

File:wieskirche boenisch okt 2003.jpg *Source*: http://en.wikipedia.org/w/index.php?title=File:Wieskirche_boenisch_okt_2003.jpg *License*: unknown *Contributors*: Bönisch

File:Bartholomae-2005.jpg *Source*: http://en.wikipedia.org/w/index.php?title=File:Bartholomae-2005.jpg *License*: unknown *Contributors*: User:Daniel FR

File:Castle Neuschwanstein.jpg *Source*: http://en.wikipedia.org/w/index.php?title=File:Castle_Neuschwanstein.jpg *License*: unknown *Contributors*: User:Softeis

File:Coat of Arms of Germany.svg *Source*: http://en.wikipedia.org/w/index.php?title=File:Coat_of_Arms_of_Germany.svg *License*: unknown *Contributors*: Wappenentwurf: Karl-Tobias Schwab (1887–1967), entworfen 1926 diese Datei: Jwnabd

File:EU-Germany.svg *Source*: http://en.wikipedia.org/w/index.php?title=File:EU-Germany.svg *License*: unknown *Contributors*: User:NuclearVacuum

File:Increase2.svg *Source*: http://en.wikipedia.org/w/index.php?title=File:Increase2.svg *License*: unknown *Contributors*: Sarang

Image:Speakerlink.svg *Source*: http://en.wikipedia.org/w/index.php?title=File:Speakerlink.svg *License*: unknown *Contributors*: Woodstone. Original uploader was Woodstone at en.wikipedia

File:Imperium Romanum Germania.png *Source*: http://en.wikipedia.org/w/index.php?title=File:Imperium_Romanum_Germania.png *License*: unknown *Contributors*: Modification · Bearbeitung · Prilaboro: D. Bachmann

File:Martin Luther by Lucas Cranach der Ältere.jpeg *Source*: http://en.wikipedia.org/w/index.php?title=File:Martin_Luther_by_Lucas_Cranach_der_Ältere.jpeg *License*: unknown *Contributors*: 1970gemini, Aavindraa, Avatar, Con2tto, DaTroll, Dr. Shaggeman, Ecummenic, FA2010, Interpretix, Krinkle, Mattes, Miya, Origamiemensch, Pomfuttge, Quadell, Rosenzweig, Sailko, Semnoz, Shakko, Túrelio, Warburg, Wst, Ævar Arnfjörð Bjarmason, 4 anonymous edits

File:Wernerprokla.jpg *Source*: http://en.wikipedia.org/w/index.php?title=File:Wernerprokla.jpg *License*: unknown *Contributors*: +МоГиЛьЩиК+, 1970gemini, 52 Pickup, Adam Cuerden, AnRo0002, Chrisportelli, Christian1985, DIREKTOR, DutchHoratius, Hajotthu, Jacklee, Jappalang, Kameraad Pjotr, Kresspahl, Leppus, Mattes, Mnmazur, Mogelzahn, Rd232, Rlevse, Shakko, Túrelio, Wuselig, Ælfgar, 7 anonymous edits

File:German Reich1.png *Source*: http://en.wikipedia.org/w/index.php?title=File:German_Reich1.png *License*: unknown *Contributors*: User:Wiggy!

File:Bundesarchiv Bild 183-S33882, Adolf Hitler retouched.jpg *Source*: http://en.wikipedia.org/w/index.php?title=File:Bundesarchiv_Bild_183-S33882,_Adolf_Hitler_retouched.jpg *License*: unknown *Contributors*: User:AlMare

File:Potsdamer Platz 1945.jpg *Source*: http://en.wikipedia.org/w/index.php?title=File:Potsdamer_Platz_1945.jpg *License*: unknown *Contributors*: Michael M. Dean (Canada. Dept. of National Defence/Ministère de la défense nationale)

File:Map-Germany-1945.svg *Source*: http://en.wikipedia.org/w/index.php?title=File:Map-Germany-1945.svg *License*: unknown *Contributors*: Original uploader was 52 Pickup at en.wikipedia

File:Thefalloftheberlinwall1989.JPG *Source*: http://en.wikipedia.org/w/index.php?title=File:Thefalloftheberlinwall1989.JPG *License*: unknown *Contributors*: Unknown photographer, Reproduction by Lear 21

File:Deutschland topo.jpg *Source*: http://en.wikipedia.org/w/index.php?title=File:Deutschland_topo.jpg *License*: unknown *Contributors*: User:Botaurus-stellaris

File:Aquila chrysaetos Flickr.jpg *Source*: http://en.wikipedia.org/w/index.php?title=File:Aquila_chrysaetos_Flickr.jpg *License*: unknown *Contributors*: MPF

File:Political System of Germany.svg *Source*: http://en.wikipedia.org/w/index.php?title=File:Political_System_of_Germany.svg *License*: unknown *Contributors*: User:111Alleskönner

File:Reichstag building Berlin view from west before sunset.jpg *Source*: http://en.wikipedia.org/w/index.php?title=File:Reichstag_building_Berlin_view_from_west_before_sunset.jpg *License*: unknown *Contributors*: User:JürgenMatern

File:2010-11-29 JoachimGauck 245.JPG *Source*: http://en.wikipedia.org/w/index.php?title=File:2010-11-29_JoachimGauck_245.JPG *License*: unknown *Contributors*: User:Pixeldost

Image:Karte Bundesrepublik Deutschland.svg *Source*: http://en.wikipedia.org/w/index.php?title=File:Karte_Bundesrepublik_Deutschland.svg *License*: unknown *Contributors*: User:David Liuzzo

File:Coat of arms of Lower Saxony.svg *Source*: http://en.wikipedia.org/w/index.php?title=File:Coat_of_arms_of_Lower_Saxony.svg *License*: unknown *Contributors*: Darwinius, Dbenbenn, Fred the Oyster, Frombenny, Gormo, Jimmy44, Lou.gruber, Ludger1961, Madden, Mattes, Maxxl2, Ollemarkeagle, Pitke, Rosenzweig, 2 anonymous edits

File:Bremen Wappen(Mittel).svg *Source*: http://en.wikipedia.org/w/index.php?title=File:Bremen_Wappen(Mittel).svg *License*: unknown *Contributors*: User:David Liuzzo

File:Coat of arms of Hamburg.svg *Source*: http://en.wikipedia.org/w/index.php?title=File:Coat_of_arms_of_Hamburg.svg *License*: unknown *Contributors*: unkown

File:Coat of arms of Mecklenburg-Western Pomerania (great).svg *Source*: http://en.wikipedia.org/w/index.php?title=File:Coat_of_arms_of_Mecklenburg-Western_Pomerania_(great).svg *License*: unknown *Contributors*: Bundesland Mecklenburg-Vorpommern

File:Wappen Sachsen-Anhalt.svg *Source*: http://en.wikipedia.org/w/index.php?title=File:Wappen_Sachsen-Anhalt.svg *License*: unknown *Contributors*: Land Sachsen-Anhalt

File:Coat of arms of Saxony.svg *Source*: http://en.wikipedia.org/w/index.php?title=File:Coat_of_arms_of_Saxony.svg *License*: unknown *Contributors*: Freistaat Sachsen; House of Wettin

File:Brandenburg Wappen.svg *Source*: http://en.wikipedia.org/w/index.php?title=File:Brandenburg_Wappen.svg *License*: unknown *Contributors*: User:David Liuzzo

File:Country symbol of Berlin color.svg *Source*: http://en.wikipedia.org/w/index.php?title=File:Country_symbol_of_Berlin_color.svg *License*: unknown *Contributors*: Wappenentwurf: diese Datei: Jwnabd

File:Coat of arms of Thuringia.svg *Source*: http://en.wikipedia.org/w/index.php?title=File:Coat_of_arms_of_Thuringia.svg *License*: unknown *Contributors*: Burts, Frombenny, Ludger1961, MB-one, Madden, Michael Sander, Mogelzahn, Palosirkka

File:Coat of arms of Hesse.svg *Source*: http://en.wikipedia.org/w/index.php?title=File:Coat_of_arms_of_Hesse.svg *License*: unknown *Contributors*: Darwinius, Ewan McTeagle, F. F. Fjodor, Frombenny, Madden, Mogelzahn, Perhelion, Rauenstein, Túrelio, 1 anonymous edits

File:Coat of arms of North Rhine-Westfalia.svg *Source*: http://en.wikipedia.org/w/index.php?title=File:Coat_of_arms_of_North_Rhine-Westfalia.svg *License*: unknown *Contributors*: Wappenentwurf (1947): Wolfgang Pagenstecher für das Land Nordrhein-Westfalen

File:Coat of arms of Rhineland-Palatinate.svg *Source*: http://en.wikipedia.org/w/index.php?title=File:Coat_of_arms_of_Rhineland-Palatinate.svg *License*: unknown *Contributors*: Land Rheinland-Pfalz

File:Landessymbol Freistaat Bayern.svg *Source*: http://en.wikipedia.org/w/index.php?title=File:Landessymbol_Freistaat_Bayern.svg *License*: unknown *Contributors*: Freistaat Bayern

File:Coat of arms of Baden-Württemberg (lesser).svg *Source*: http://en.wikipedia.org/w/index.php?title=File:Coat_of_arms_of_Baden-Württemberg_(lesser).svg *License*: unknown *Contributors*: AndreasPraefcke, BrightRaven, Darwinius, Ewan McTeagle, Frombenny, Gryffindor, Jed, Madden, Milgesch, Rtc, Savh, Ssch, 2 anonymous edits

File:Wappen des Saarlands.svg *Source*: http://en.wikipedia.org/w/index.php?title=File:Wappen_des_Saarlands.svg *License*: unknown *Contributors*: Entwurf vom Landesarchiv Saarbrücken für das Saarland

File:Coat of arms of Schleswig-Holstein.svg *Source*: http://en.wikipedia.org/w/index.php?title=File:Coat_of_arms_of_Schleswig-Holstein.svg *License*: unknown *Contributors*: Burts, Drbashir117, Frombenny, Mogelzahn, Rosenzweig, Shizhao, 1 anonymous edits

File:33rdG8Leaders.jpg *Source*: http://en.wikipedia.org/w/index.php?title=File:33rdG8Leaders.jpg *License*: unknown *Contributors*: White House photo by Eric Draper

File:SLS AMG Roadster.jpg *Source*: http://en.wikipedia.org/w/index.php?title=File:SLS_AMG_Roadster.jpg *License*: unknown *Contributors*: User:Ritchyblack

File:Eurozone.svg *Source*: http://en.wikipedia.org/w/index.php?title=File:Eurozone.svg *License*: unknown *Contributors*: User:Glentamara

File:InterCityExpress Frankfurt.jpg *Source*: http://en.wikipedia.org/w/index.php?title=File:InterCityExpress_Frankfurt.jpg *License*: unknown *Contributors*: User:Der Wolf im Wald

File:Einstein 1921 by F Schmutzer.jpg *Source*: http://en.wikipedia.org/w/index.php?title=File:Einstein_1921_by_F_Schmutzer.jpg *License*: unknown *Contributors*: Calliopejen1, DIREKTOR, Frank C. Müller, Hemulen, Lobo, Quibik, Raeky, Андрей Романенко

File:Population Pyramid of Germany in 2012.png *Source*: http://en.wikipedia.org/w/index.php?title=File:Population_Pyramid_of_Germany_in_2012.png *License*: unknown *Contributors*: International Futures (IFs)

File:Kdom.jpg *Source*: http://en.wikipedia.org/w/index.php?title=File:Kdom.jpg *License*: unknown *Contributors*: ger1axg

File:Heidelberg Universitätsbibliothek 2003.jpg *Source*: http://en.wikipedia.org/w/index.php?title=File:Heidelberg_Universitätsbibliothek_2003.jpg *License*: unknown *Contributors*: Jan Beckendorf

File:Beethoven.jpg *Source*: http://en.wikipedia.org/w/index.php?title=File:Beethoven.jpg *License*: unknown *Contributors*: ADGE, Aaron1a12, Aavindraa, Butko, Codeispoetry, CommonsDelinker, Herbythyme, Hoo man, Hotshot977, Howcheng, Ies, Interpretix, Jappalang, Johney, Jusjih, Leonce49, Maclemo, Mattes, Michael Bednarek, Neukoln, Pixel ;-), Shakko, Sonphan10, Thuresson, אמסיסה123, 5 anonymous edits

File:Grimm.jpg *Source*: http://en.wikipedia.org/w/index.php?title=File:Grimm.jpg *License*: unknown *Contributors*: AndreasPraefcke, Anne-Sophie Ofrim, Boo-Boo Baroo, Bukk, Gabor, Hannah, Kresspahl, Léna, Mattes, Nillerdk, Origamiemensch, Phrood, Reinhard Dietrich, Shakko, Yuval Y, 5 anonymous edits

File:Black Forest gateau.jpg *Source*: http://en.wikipedia.org/w/index.php?title=File:Black_Forest_gateau.jpg *License*: unknown *Contributors*: Mikelo

File:Singal Iduna Park with a new sign.jpg *Source*: http://en.wikipedia.org/w/index.php?title=File:Singal_Iduna_Park_with_a_new_sign.jpg *License*: unknown *Contributors*: User:DR

FSC
www.fsc.org
MIX
Papier aus verantwortungsvollen Quellen
Paper from responsible sources
FSC® C105338

Printed by Books on Demand GmbH, Norderstedt / Germany